Prompt Engineering for Busy Marketers

Prompt Engineering for Busy Marketers

Turn AI into your Marketing Assistant with Plug-And-Play Prompts that Get Results

Author: Viki Zangari

Publisher: Zangari Publishing

Western Australia 2025

This book is a work of nonfiction. While the techniques, examples, and strategies reflect real marketing practices and personal experiences, readers should evaluate and adapt the content according to their own business, audience, and circumstances. The author and publisher disclaim liability for any loss or damages resulting from the use of this book. Portions of this book were developed through collaborative work with AI tools. All final content has been human-reviewed and shaped with care.

Published by Zangari Publishing Western Australia

ISBN: 978-1-7643993-1-9

Cover design: Author Services Australia

Interior design and formatting: by "Zangari Publishing, designed in Atticus"

Website: www.zangaripublishing.com

e-mail: zangaripublishing@gmail.com

Dedication

This book is dedicated to the professionals who juggle deadlines, client expectations, shifting priorities, meetings, metrics, emails, and the weight of constant adaptation.

To the marketers who write headlines between responsibilities, refine ideas while travelling, review campaigns during brief pockets of quiet, and still manage to keep moving forward — this work honours you.

My hope is that this book becomes your companion — a guide you can return to when you feel scattered, pressed for time, or unsure how to take the next step. May it lighten your workload, sharpen your thinking, and offer a sense of steadiness on the days when you need it most.

To all the doers, thinkers, creators, educators, leaders, small-business owners, and quiet achievers who keep showing up — even when the world demands more than feels fair... This book is for you.

May you find ease in the process, confidence in your voice, and reassurance in knowing that progress is always possible — even in the busiest seasons of your life.

To the people who work during lunch breaks, steal moments of creativity between school runs, think about campaign ideas in the car, and somehow find a spark of energy left at the end of the day — I see you. I understand you. And this book was written with you in mind. ——-Viki

Contents

Forward

Marketing has never lacked tools. What it has consistently lacked is sustainability.

Over the last decade, marketers have been asked to do more with less: more platforms, more content, more responsiveness, more visibility—often with fewer resources, tighter timelines, and increasing cognitive load. At the same time, the industry has normalised a pace of work that assumes uninterrupted focus, constant availability, and unbroken creative output. For many professionals, that model is not only unrealistic; it is actively corrosive. Prompt engineering, artificial intelligence, and automation are often presented as solutions to speed and scale. In practice, they have created a second problem: complexity layered on top of already strained workflows. Many marketers now find themselves responsible not only for strategy and execution, but also for learning, managing, and optimising tools that were supposed to make their work easier.

What is missing from much of the conversation is a discussion about how marketing work actually fits inside a busy life.

This is where *Prompt Engineering for Busy Marketers* stands apart.

Rather than promising shortcuts, growth hacks, or instant leverage, this book does something far more valuable. It treats clarity, structure, and rhythm as professional competencies. It recognises that consistency is not a motivational issue, that resistance is not a personal flaw, and that productivity is not measured solely by output volume. Instead, it reframes marketing as a practice that must be designed to function under real-world conditions.

Sustainable performance has never come from intensity alone. It comes from systems that reduce unnecessary decision-making, foundations that preserve cognitive energy, and workflows that adapt to fluctuation rather than collapse under it. This book reflects that understanding.

What you will not find here is a technical manual on artificial intelligence, nor a collection of generic prompts detached from context. You will not be asked to overhaul your entire set up overnight or adopt a version of marketing that ignores your capacity.

Instead, you are guided through a way of working that respects attention, reduces friction, and restores agency to the marketer. The integration of prompt engineering is particularly noteworthy. Rather than positioning AI as a replacement for thinking or creativity, it is presented as a stabilising force—one that supports clarity, lowers the energy cost of starting, and absorbs cognitive overhead when energy is limited.

This is a mature, professional stance that reflects real-world use rather than theoretical possibility.

Equally important is the emphasis on rhythm.

In high-performing teams, rhythm is often the invisible factor behind consistency. It is the difference between a system that survives disruption and one that collapses the moment conditions change. By focusing on rhythm rather than routine, this book offers a model that accommodates variability while preserving momentum. That understanding alone will resonate with marketers who have struggled to maintain output without burning out.

Throughout these pages, you will encounter practical structures, clear architecture, and prompts that are immediately usable. But more than that, you will come across a way of thinking about marketing work that feels grounded and credible. The advice here does not assume perfect conditions. It assumes competing priorities, interrupted time, emotional labour, and fluctuating capacity—and it works within those constraints rather than against them.

I would encourage readers to approach this book not as something to rush through, but as something to work with. The value is not in consuming every page at speed, but in noticing where friction begins to ease, where clarity replaces pressure, and where consistency becomes less effortful.

The examples introduced here are designed to be revisited, adapted, and integrated over time. Marketing is ultimately a long game. Trust is built gradually. Voice is refined through repetition. Systems earn their value by surviving real conditions, not ideal ones.

This book recognizes that. For busy marketers seeking a way to work with intelligence, structure, and sustainability—without sacrificing quality or credibility—*Prompt Engineering for Busy Marketers* offers a thoughtful and timely contribution to the field.

My hope is that this book becomes a trusted collaborator in your daily workflow. A quiet mentor sitting beside you. A source of structure when your mind feels scattered. As you move through the pages ahead, take what speaks to you, adapt it to your world, and let it evolve with you. AI is not here to replace your voice, your strategy, or your intuition. It's here to sharpen them.

Thank you for choosing to step into this journey with me. Let's unlock a way of working that helps you feel more organized, more confident, and far less overwhelmed

Introduction

Prompt Engineering for Busy Marketers

A practical, confidence-building guide for marketers who need results — fast.

Whether you manage campaigns, create content, or support a small business, the pressure to produce more in less time is constant. That's where prompt engineering becomes a competitive edge rather than a technical skill.

Inside, you'll learn how to use AI as a strategic partner — to help streamline your workflow, accelerate your ideas, and remove the mental bottlenecks that slow you down.

The engineered prompts in these pages are ready to use, highly adaptable, and grounded in real marketing scenarios: writing social content, optimizing campaigns, building brand messaging, repurposing assets, and reducing the time it takes to go from idea to execution. Each chapter shows you how to move faster, think clearer, and maintain a professional standard without sacrificing quality.

This is not theory. It's a system — one you can apply from the moment you turn the page. If you're ready to reduce your workload, sharpen your message, and create with confidence, you're in the right place. Turn the page, and let's begin. —Viki

Chapter One

Time-Saving Prompts for Everyday Marketing Tasks

Section One

Why Marketers Need Prompt Efficiency

Marketing today moves at a pace most people never get to see — and yet, busy marketers are expected to keep up with all of it. You're writing content, planning campaigns, checking analytics, answering messages, posting consistently, creating visuals, and somehow still trying to stay inspired. Prompt efficiency isn't about rushing your work. It's about reducing mental load, simplifying decisions, and saving precious time so you can focus on what matters most — strategy, creativity, and connection with your audience. Many marketers struggle not because they lack skill, but because they're juggling too many tasks at once. I understand this deeply, and I want to make your load lighter, not heavier. Prompt efficiency is often misunderstood. It's not about cutting corners, lowering standards, or automating creativity out of your work. It's a professional skill — one that experienced marketers develop deliberately over time. In traditional marketing workflows, clarity comes after long planning sessions, multiple drafts, and countless revisions. Prompt efficiency reverses that order. When you give AI clear direction upfront, clarity comes first — and execution becomes faster, lighter, and more consistent.

When prompts are vague, the output feels vague. When prompts are structured, the output becomes usable. Over time, this changes how you approach marketing altogether. You stop staring at blank pages. Or second-guessing every sentence. You stop burning energy on decisions that don't justify it. Prompt efficiency gives you a repeatable starting point — and in a busy marketing role, that starting point is everything.

Here's why prompt efficiency is essential

1. It reduces overwhelm when your brain is already full. When your mind is scattered or tired, prompting may feel harder than it should. Efficient prompts help remove the guesswork.

Inefficient prompt: "Write a post about my business." **Use: the following**

PROMPT: Act as a senior marketer and create **3** Facebook posts for a small, local service-based business. The tone must be warm, trustworthy, and community focused. Each post must be under 50 words and written to encourage new clients to book or enquire. Please include: A clear hook in the first sentence. One customer benefit (comfort, reliability, timesaving, peace of mind, etc.). A gentle call-to-action suitable for Facebook (Book now, Message us, Learn more). A platform-ready format (short sentences, easy to skim, natural voice). Business Type: [Insert business type, e.g., cleaning service] Target Audience: [Insert audience] Offer or Key Message: [Insert any promotion or focus] Brand Voice: [Insert brand voice]

One clear prompt replaces 20 minutes of thinking

***2.* It gives you consistency – even on tough days.** Some days you're inspired. Other days, you're mentally done before you start. Prompt efficiency ensures your content stays high-quality regardless of your energy.

Example: A structured prompt like

PROMPT: Rewrite the caption below in a friendlier, more engaging tone while keeping the original message intact. Then create **t**wo shorter alternative versions optimized for social media. Please include Version 1: A polished, friendlier rewrite (same meaning, warmer tone). Version 2: A short version under 20–25 words. Version 3: A very short version under 10–12 words, suitable for fast-scrolling feeds. Ensure all versions: Maintain brand voice Keep the message clear Use simple, audience-friendly language Original Caption: [Paste your caption **here]** Brand Voice: [Insert brand voice] Audience: [Insert audience]

3. It saves hours each week – without sacrificing quality. Most marketers spend too long drafting and redrafting. Efficient prompts help you move faster *and* smarter.

Realistic time-saving Prompts: Write a blog outline in 10 seconds

***PROMPT**: Create a clear, SEO-friendly blog outline in under 10 seconds. Follow this structure: Craft a compelling working title based on my topic. Identify the target audience and the problem this blog addresses. Provide 4-6 scannable section headings that follow a logical flow. Include 1-2 bullet points under each section explaining what should be covered. Suggest a strong call-to-action that aligns with typical marketing goals. Topic: [Insert your topic] Audience: [Insert audience] Goal of the blog: [Traffic / Leads / Education / Authority]

When This Type of Prompt Is Most Useful; Use structured prompts like this when: you're short on time but still need polished output: you know what you want, but not how to phrase it: your energy is low and decision-making feels heavy: you want consistent quality without starting from scratch: These prompts are especially powerful during content batching days: campaign launches: busy weeks where marketing must stay visible, even if creativity feels flat: The goal is momentum you can sustain.

Turn one blog into 10 social posts instantly

PROMPT; Repurpose the following blog into 10 high-performing social media posts. Use a mix of formats to maximize reach and engagement. Please include: A variety of post styles: 2 educational tips 2 problem/solution posts 2 value statements or insights 2 engagement questions 2 short promotional posts Platform optimization: Tailor the tone and length for: Facebook Instagram LinkedIn Marketing essentials: Add a hook for each post Include a clear call-to-action Suggest 3-5 relevant hashtags for each platform Blog Content: [Paste your blog here] Brand Voice: [Insert brand voice] Primary Goal: [Engagement / Traffic / Lead generation]

Generate ad variations without starting from scratch

PROMPT; Create multiple high-performing ad variations based on the original ad copy below. Keep the core message intact but produce fresh angles that marketers can A/B test. Please provide: 5 headline variations Mix of benefit-driven, curiosity-driven, and urgency-driven options 5 primary text variations Rewrite using different tones: authoritative, friendly, concise, aspirational, and problem/solution. 3 call-to-action variations Optimized for conversions. Audience angle shifts: Create 3 ad versions that speak to different audience segments. Example segments: beginners, busy professionals, budget-conscious buyers, or advanced users. Platform formatting: Tailor the final versions so they can be used on: Facebook Instagram LinkedIn Original Ad Copy: [Paste your ad **here]** Target Audience: [Insert audience] Goal of the Ad: [Traffic / Conversions / Lead Generation / Awareness]

Repurpose content for multiple platforms in minutes

PROMPT: Repurpose the following piece of content into platform-ready formats for Facebook, Instagram, LinkedIn, and TikTok. Maintain message consistency while optimizing tone, structure, and length for each platform. Please include Platform-specific versions: Facebook: short paragraph + CTA Instagram: caption with hook + line breaks + 5–8 hashtags LinkedIn: polished, informative version with a professional angle TikTok: 15–20 second script with a clear hook and call-to-action Brand alignment: Match the content to the brand voice I provide. Audience targeting: Highlight what matters most to the intended audience **segment.** Marketing essentials: Add hooks, micro-stories, or insights to increase engagement. Suggest 3–5 hashtags per platform (except LinkedIn where optional). Original Content: [Paste your content **here]** Brand Voice: [Insert brand voice] Target Audience: [Insert audience] Primary Goal: [Awareness / Engagement / Traffic / Leads] Saving time isn't a luxury. It's a survival skill in marketing.

Prompt Efficiency Reduces Decision Fatigue

One of the most underestimated drains on marketers is decision fatigue. Every caption, headline, email, or campaign requires dozens of micro-decisions — tone, structure, length, goal, platform, and audience. When you prompt efficiently, you pre-decide these elements once and reuse them repeatedly. This reduces cognitive load and allows you to spend your mental energy where it matters most: positioning, messaging, and strategy. Instead of asking yourself: "How should this sound?": "What angle should I take?": "Is this good enough?" You're simply choosing from outputs that already align with your standards. Over time, this creates a calmer workflow. Marketing stops feeling like constant problem-solving and starts feeling like execution. That shift alone can reclaim hours every week.

***4.* It helps you think more clearly**. AI becomes a thinking partner, not just a writing tool. When you prompt efficiently, AI can: organize your ideas refine your strategy help you see what's missing guide you when you feel stuck

Brainstorm high value Content Ideas

PROMPT: Brainstorm 10 high-value content ideas for a small service-based business. Priorities themes that build customer trust, highlight visible transformations, and reinforce a strong self-care or customer-benefit message. Please include: A mix of content types: Educational posts Before/after transformations Customer trust builders Service spotlights Self-care or lifestyle value posts Marketing structure: Each idea must include a headline, a one-sentence description, and the angle (trust / transformation / self-care). Platform flexibility: Ideas should be suitable for repurposing across Instagram, Facebook, TikTok, and websites. Optional enhancement: Suggest one visual concept per idea (photo, short video, carousel, reel, testimonial graphic). Business Type: [Insert business type—e.g., beauty salon, barber, spa, clinic] **T**arget

Audience: [Insert audience] Brand Voice: [Insert brand voice]

This gives clarity you can build on immediately

5. It creates breathing room in your busy week. Prompt efficiency gives you space by removing unnecessary steps. It lightens your workload so you can: work smarter, not harder regain creative confidence avoid burnout balance work with real life.

6. It turns AI into a true partner instead of a guessing game. AI becomes: your editor your strategist your idea generator your clarity booster your time-saver. This book will teach you how to build that partnership, so your marketing doesn't rely on luck — it relies on skill.

A quick before-and-after example—

Inefficient prompt: "Write an email for my business"

Why Structured Prompts Scale Better

The real value of efficient prompts isn't just speed — it's scalability. When you use clear prompt structures, you can: reuse them across platforms: adapt them for new campaigns: delegate them to team members or VAs: return to them months later without relearning your process. This is how solo marketers work like teams and how busy professionals maintain consistency without burnout. The work compounds instead of restarting each time. Prompt efficiency turns AI into infrastructure — not a novelty.

Efficient prompt: REVISED PROMPT: Act as an email marketing expert and write a short welcome email for new subscribers to a small product-based business. The tone must be warm, simple, and friendly, and the goal is to make subscribers excited to explore the product range. Keep the email under 120 words and include one clear call-to-action. Please include: A friendly greeting that feels personal and welcoming. A one-sentence introduction to what the business offers. A short value statement (what makes the products special, handmade, or unique). A clear, single CTA encouraging readers to explore or shop. A closing line that reinforces warmth and community. Business Type: [Insert business type—e.g., handmade candle business] Brand Voice: [Insert brand voice] Audience: [Insert audience]

The difference? One wastes time. The other saves it.

Why this matters for you! I care about helping you gain clarity because I know what it feels like to struggle through tasks that should be simple. I know what it's like to juggle responsibilities, feel mentally overloaded, and still want to do good work. Prompt efficiency is about giving yourself a system that supports you — especially on the days when life is busy and energy is limited.

Section Two

From Understanding to Execution

Understanding why prompt efficiency matters is important — but its real value shows up when you apply it. The following prompt sets are designed for everyday marketing tasks, they're practical, reusable, and intentionally structured to reduce effort while improving results. Each set gives you: a clear role for AI: defined output expectations: built-in constraints that protect quality. You can copy them as-is, adapt them to your voice, or evolve them over time. What matters is that they give you a dependable starting point.

Now that you are aware why prompt efficiency matters, let's move into the practical side — the prompts you can start using immediately to speed up your marketing, reduce decision fatigue, and stay consistent week after week. They're simple, repeatable, and aligned with the real rhythms of modern marketing, these prompts are built to make your life easier. Each example includes a short explanation, so you understand *why* it works — not just how to copy and paste it.

PROMPT Set 1 — Writing a Weekly Newsletter: Newsletters take time, especially if you're starting from a blank page. This prompt gives you structure, tone, and flow without losing your personality. Use this when: You want a human, relatable email that builds trust without sounding salesy.

PROMPT: Act as an email copywriter for a values-based business. Create a 3-paragraph newsletter that: Paragraph 1: Opens with a relatable insight, story, or everyday moment that connects emotionally with the audience. Paragraph 2: Shares one practical tip aligned with the brand's values and helpful for the reader's daily life or goals. Paragraph 3: Closes with a soft call-to-action that feels supportive rather than sales-driven. Please ensure: Tone is warm, grounded, and aligned with the business values. Writing is concise but meaningful. The CTA feels natural, optional, and empowering. Business Type / Niche: [Insert business type] Brand Values: [Insert values] Audience: [Insert audience] Topic of the Newsletter: [Insert topic]

Why it works: Gives AI a clear role Defines the structure Ensures a natural beginning, middle, end. Avoids pushy or robotic language

PROMPT Set 2—Creating Instagram Captions: Captions need to feel conversational, simple, and engaging – especially on busy days. This prompt helps you turn longer content into short, shareable posts. Use this when: You have a blog, email, or long message and want a quick social version.

PROMPT; Act as a brand voice strategist and transform the blog excerpt below into a casual, conversational Instagram caption. Please include: A relatable hook in the first sentence to pull readers into the caption. A simplified, friendly version of the blog excerpt that fits Instagram's casual tone. One engaging question at the end to increase comments and interaction. Ensure the caption maintains the brand's voice and feels natural for social media. Blog Excerpt: [Paste excerpt **here]** Brand Voice: [Insert brand voice] Audience: [Insert audience]

Why it works: Opens with a hook (essential for IA) Keeps the tone relaxed Encourages engagement with a question Saves you from rewriting everything manually

PROMPT Set 3 – Repurposing a Blog Post for LinkedIn: LinkedIn requires a slightly more professional tone but still needs clarity and personality. This prompt blends both. Use this when: You want to expand your reach by turning long-form content into a polished LinkedIn post.

PROMPT; Summarize the blog post below into a 150-word LinkedIn post. Maintain a confident, professional tone suitable for industry audiences. Highlight one key insight that positions the brand or author as an authority. Please include: A strong opening line that captures attention in the feed. A concise synthesis of the main idea from the blog post. One standout insight that would resonate with professionals. A clear closing statement that reinforces credibility or invites further reflection. Blog Post: [Paste blog post here] Industry / Audience: [Insert audience] Brand Voice: [Insert brand voice]

Why it works: Sets a clear word limit Preserves your authority and message Makes repurposing fast and consistent

PROMPT Set 4 – Responding to a Customer Question: Customer service replies need to be clear, warm, and calm – especially when a customer is confused or unsure. This prompt helps you respond professionally and empathetically in seconds. Use this when: You need a quick, polished response without sounding rushed.

PROMPT; Act as a customer support specialist and write a clear, empathetic reply to the customer question below. Keep the response under 100 words, maintain a supportive tone, and ensure the customer feels heard and guided. Please include: A brief acknowledgment of the customer's concern or question. A simple, direct explanation or solution. One reassuring line that reinforces trust. Optional next step (link, resource, or clarification request). Customer Question: [Paste the customer's question here] Brand Voice: [Insert brand voice] Product Type: [Insert product]

Why it works: Keeps the reply short and readable Maintains professionalism Reduces emotional strain when replying under pressure

Why These Prompts Matter: These prompt structures aren't random. They reflect the core marketing activities most professionals return to every single week: writing newsletters, posting on social media, repurposing content, responding to customers. By using consistent prompt templates, you remove friction from your workflow and save mental energy for the creative or strategic tasks that genuinely need your attention. My goal is to help you build a system that feels supportive, not overwhelming — a toolkit you can rely on when life is busy, and your energy is limited.

Section 3

A Simple First Win — Use One Prompt Today

You don't need to master every strategy in this book today. You don't need to understand all the tools, frameworks, or techniques at once. In fact, trying to learn everything too quickly is what overwhelms most marketers in the first place. Prompting works best when you build confidence through small, repeatable successes.

Why this works: When you try one well-structured prompt, three things happen: You see immediate progress. A task that felt heavy suddenly feels manageable. You experience the relief of clarity. Instead of pushing through mental fog, you gain support and structure. Your confidence rises naturally. Not from hype or pressure, but from experiencing a real, practical win. Small wins compound more quickly than people realize. One simple success can shift your entire relationship with marketing — because it shows you that you're not alone in the process anymore. You have a partner. You have support. You have a way forward.

Turn a blog snippet into a social post

PROMPT: Turn the blog snippet below into a high-performing social media post. Maintain the core message while adapting the tone and structure for fast-scrolling platforms. Please include: A strong hook in the first sentence to capture attention. A concise, audience-friendly rewrite of the snippet. A clear takeaway or insight that adds value. A soft call-to-action appropriate for social channels. Optional: Suggest 3–5 relevant hashtags. Blog Snippet: [Paste snippet here] Platform: [Instagram / Facebook / LinkedIn / TikTok text overlay] Brand Voice: [Insert brand voice] Audience: [Insert audience]

Draft a customer reply in under 100 words

PROMPT: Draft a concise, customer-friendly reply to the message below in under 100 words. The response should be clear, polite, and solution-focused, ensuring the customer feels acknowledged and supported. Please include: A brief acknowledgment of their message or concern. A simple explanation or answer in plain language. One reassuring or helpful line to build trust. A clear next step (if needed). Customer Message: [Paste customer message here] Brand Voice: [Insert brand voice] Product or Service: [Insert product/service]

Convert a long idea into a short caption

PROMPT: Convert the long idea below into a short, punchy caption suitable for social media. Maintain the core message but simplify the language and shorten the structure. Please include Version 1: A concise caption under 25–30 words. Version 2: An ultra-short caption under 12–15 words. Ensure both versions use a clear hook, friendly tone, and brand-aligned voice. Long Idea: [Paste idea here] Brand Voice: [Insert brand voice] Platform: [Instagram / Facebook / LinkedIn]

Summarize a blog for LinkedIn

PROMPT: Summarize the blog below into a clear, professional LinkedIn post. Maintain an authoritative tone suitable for industry readers while keeping the message concise and insight driven. Please include: A strong opening line that captures attention. A concise summary of the blog's core message. One key insight or takeaway that positions the brand or author as a thought leader. A short closing line that encourages reflection, discussion, or connection. Blog Content: [Paste blog here] Industry / Audience: [Insert audience] Brand Voice: [Insert brand voice]

Generate three newsletter intros to choose from

PROMPT: Generate three distinct newsletter introduction options for the topic provided below. Each intro should be engaging, concise, and aligned with the brand voice, giving me different tones to choose from. Please include Version 1: Warm and conversational Version 2: Professional and insight-driven Version 3: Energetic and high-impact Ensure each intro: Hooks the reader in the first sentence Sets up the theme of the newsletter Remains under 60–80 words Newsletter Topic: [Insert topic] Brand Voice: [Insert brand voice] Audience: [Insert audience] These are quick, low-pressure tasks — but each one gives you a feeling of movement, Confidence doesn't come from thinking about the work. It comes from doing the work — even in the smallest way.

In the next chapter. I'll show you how to build a prompting mindset: a way of thinking that turns AI from a tool into a workflow you can rely on anytime your ideas feel tangled or your schedule becomes overwhelming. Try a prompt. Take one clear action.

Closing Reflection

You don't need to overhaul your entire marketing system or memorise complex prompts. What matters is noticing the difference when clarity replaces friction — and how much lighter the work feels when guidance replaces guesswork.

Forward into chapter two

Chapter Two shifts from prompts to mindset: how you think, how you approach your work, and how AI can support you without adding pressure. You'll begin building a way of working that reduces mental load and holds steady even when life is full. You've started — and that matters

Chapter Two

The Prompt Mindset for Marketers Balancing Real-Life Demands

Section One

What most marketers actually need is not a hack — it's support. Support that is flexible. Support that adapts to your energy. Support that works around real-life demands instead of demanding more from you. That's how I came to AI. Not as a tech expert. Not as someone with extra time. I came in with questions that mattered to my life:

Can this Advanced Intelligence help me keep my creativity alive?

Can this Advanced Intelligence help me work faster without losing my voice?

Can this Advanced Intelligence help me stay productive on the days I feel drained?

Can Advanced Intelligence fit between caring duties, housework, and planting tomatoes?

I didn't need theory. I needed a tool that understood real-world limits. So, I tested a variety of systems, Claude, DeepSeek, Jasper, Midjourney, Stable Diffusion and more. I wanted to understand how these tools "thought," how they responded, and what made them useful. But it was ChatGPT that stayed with me. And not because it was perfect — but because it met me where I was. During those first days, I would prompt ChatGPT but it often did not respond. It would start "thinking" and then stop, again and again, until my confidence began to wilt. When I started to believe that advanced intelligence might be too advanced for me, something shifted. I stopped overthinking and simply began typing, I was done and needed to vent. I introduced myself, explained that I was learning how ChatGPT works and exploring what it could do.

We were having a conversation, like speaking to a mate, I had no filters, I moaned, winged and opened up with my needs and complaints. I asked AI could it help me choose a niche for a book. When I read the reply, ChatGPT had introduced itself and offered me three names to choose from, I chose "Sage." In that moment, advanced intelligence spoke back to me, I jumped out of my seat, laughing and crying with excitement and relief. It was responding like a mentor. By the end of the week, I realised I had not only gained a partner—I had gained a team capable of picking up the thread of my day, even when my own thoughts were scattered. **AI listened**. The more of myself I shared, the more it understood my needs, my flexible routines, and the limits of my technical ability. This advanced intelligence adapted to my personality.

That's when I realized something important: Prompts aren't shortcuts. They're conversations. They help you think better, faster, and more clearly. Whether you're launching a product, planning a campaign, creating content, or repurposing what you already have — prompting can lighten your load without compromising your voice. You don't need to overhaul your workflow. You just need a mindset that treats prompting as a process, not a trick.

Section Two

Prompting That Clears the Mental Load

When your mind is congested, creativity becomes harder, not easier. That's why prompting isn't just a skill — it's a *relief*. A quiet support system that brings clarity right when you need it most. You don't need more tools. You need clarity at the moment you sit down to work. And that clarity begins with intentional engineered prompting, not guessing. Here's what I learned: AI performs the way a great assistant does: Give it direction, and it delivers. Give it pressure, and it falters. When you prompt with clarity, the mental noise drops. You stop second-guessing yourself, and you get results you can actually use — without the endless rewrites.

That's where the **RIC Prompting Framework** comes in. It's the simplest, most reliable formula I've used, and it's perfect for marketers balancing real-life demands.

The RIC Prompting Framework

Role – Input – Context A simple structure that turns scattered thoughts into usable, clear content. Whether you're writing an email, planning a post, or clarifying an idea, RIC keeps you focused and saves you from overthinking. Below is a deeper, strengthened explanation of *how* to use RIC — with step-by-step guidance you can apply immediately.

Step-by-Step: How RIC Works

STEP 1 – ROLE: Tell the AI who it should "be. "This sets the tone, level of expertise, and depth of response.

Why it matters: If you don't assign a role, AI guesses — usually wrong If you *do* assign a role, AI aligns itself with the standard you want.

Examples of strong roles: Act as a senior social media strategist." "Act as a customer support specialist." "Act as a brand voice consultant." Act as a marketing copywriter for small businesses.

PROMPT: Act as a content strategist for small businesses and develop a clear, actionable content plan based on the details provided below. Please include: A 3–5 point content strategy focus (visibility, trust-building, education, local awareness, lead generation). **Five content ideas** tailored to the business type and audience. Platform recommendations (which platforms and why). Tone + messaging guidance to keep content consistent with the brand voice. One quick-win idea the business can publish immediately. Business Type: [Insert business type] Target Audience: [Insert audience] **Brand** Voice: [Insert brand voice] Primary Goal: [Awareness / Engagement / Leads / Sales]"

This instantly elevates the level of thinking.

STEP 2: INPUT: Tell AI what you're working on. This is the "raw material" the AI needs. It can be a draft, an idea, a short description, or even a bullet point.

Why it matters: AI needs something concrete to work with. No input = generic output.

Examples of helpful input: A blog topic. A product description. (A goal (like building trust or increasing sign-ups). A short messy paragraph you want rewritten

PROMPT; Create a promotional message for a downloadable checklist designed to help busy business owners organize their weekly content in 30 minutes or less. The copy should clearly communicate the benefit, highlight time savings, and encourage the audience to download. Please include: A strong hook that speaks to overwhelmed or time-poor business owners. One sentence on what the checklist does and why it's useful. One benefit-focused line (clarity, structure, consistency, reduced stress). A clear call-to-action to download the checklist. Optional: Provide one long version (45–60 words) and one short version (20–25 words). Audience: [Insert target audience] Brand Voice: [Insert brand voice] Platform: [Instagram / Facebook / LinkedIn / Email]

This gives the model something meaningful to shape.

STEP 3: CONTEXT: Add tone, audience, and format. This is where you refine the result: Who is this for? How should it sound? What format do you want?

Why it matters: Context prevents guesswork and ensures the output matches your brand.

Examples of context elements

Tone: friendly, confident, simple Format: caption, email, headline Audience: busy families, first-time buyers, small business owners Length: 50 words, 3 paragraphs, short and punchy

PROMPT; Write a short email that introduces the checklist described below. The tone must be practical, encouraging, and jargon-free, speaking directly to business owners who manage most tasks on their own. Please include: A simple, relatable opening line that acknowledges their busy workload. A brief explanation of what the checklist is and how it helps. One clear benefit (clarity, structure, time saved, reduced overwhelm). A gentle call-to-action to download or open the checklist. Keep the email under 120 words, easy to read, and supportive. Checklist Description: [Insert checklist details] Audience: Business owners managing things independently Brand Voice: [Insert brand voice]

FULL RIC Example (Put Together)

PROMPT: Act as a content strategist for small businesses and write a **short, encouraging email** introducing a downloadable checklist that helps business owners plan their weekly content in 30 minutes or less. The tone must be clear, friendly, supportive, and jargon-free. Please include: A relatable opening line that acknowledges how busy small business owners are. A simple explanation of what the checklist is and how it works. One benefit-focused sentence (clarity, structure, reduced overwhelm, time saved). A gentle call-to-action encouraging them to download or open the checklist. Keep the email under 120 words and easy to skim. Checklist Details: [Insert description] Brand Voice: [Insert brand voice] Audience: Small business owners managing most tasks themselves

This one prompt saves 20–30 minutes of thinking and delivers a professional output instantly.

Try It Yourself Template:

Here's a fill-in-the-blanks version you can use for ANY marketing task:

PROMPT; Act as a [type of professional — e.g., senior copywriter, brand strategist, social media manager, email marketer]. I'm working on [brief description of your idea, offer, campaign, content, or product]. Write [the type of content you need — e.g., a caption, email, headline, value proposition, landing page section] in a [tone/style — e.g., friendly, persuasive, professional, bold] for [specific audience – e.g., busy business owners, new subscribers, local customers, online shoppers]. Keep it [length or clarity requirement – e.g., under 50 words, concise, jargon-free, high-impact]. If you want, I can now create a bonus "advanced marketer version" of this universal prompt with additional strategic levers (hooks, CTAs, segmentation, positioning, etc.).

This removes guesswork and gives you a reliable starting point every time.

BONUS "Advanced Marketer Version

PROMPT: Act as a [type of marketing professional – e.g., senior copywriter, brand strategist, performance marketer, content director]. I'm working on [brief description of the idea, offer, product, or campaign]. Create [type of content needed – e.g., email, headline set, landing page section, ad variations, caption, script] using the following strategic requirements: Audience & Intent: Target [specific audience segment] with [intent – awareness, consideration, conversion, retention]. Messaging Angle: Emphasise [core angle – benefit, problem/solution, value, urgency, aspiration, transformation, authority]. Brand Voice: Write in a [tone/style – e.g., confident, warm, bold, minimalist, conversational] consistent with [brand personality traits]. Performance Goal: Optimise the content for [goal – engagement, clicks, sign-ups, sales, shares, watch time]. Format Requirements: Keep it [length requirement], make it [clarity requirement – scannable, punchy, jargon-free], and include [any required elements – CTA, hook, bullet points, variations, hashtag set, segmentation options]. Context or Source Material: [Paste notes, draft, transcript, or raw idea here]

3 Usable Prompts You Can Use Today

1. For Product Descriptions

PROMPT; Act as an e-commerce copywriter. Here are the key features of my product: [paste features]. Write a friendly, benefit-focused product description under 120 words, tailored for busy customers who skim-read. Highlight the problem this product solves and make the benefits easy to grasp at a glance. Please include: A clear opening line that speaks directly to the customer's pain point. A short, scannable description converting features into real-life benefits. One sentence showing how the product makes life easier, faster, or better. A gentle, confidence-building closing line that supports decision-making. Audience: Time-restricted shoppers Brand Voice: [Insert brand voice]

2. For Social Media Engagement

PROMPT: Act as a social media strategist. Turn the idea below into three short Instagram captions. Each caption must include: A relatable hook in the first line, A casual, real, helpful tone, Quick, practical value for everyday customers, A question at the end to drive engagement. Keep each caption short, easy to follow, and mobile-friendly. Idea: [Insert idea] Audience: Everyday customers who want fast, practical value Brand Voice: [Insert brand voice]

3. For Quick Clarity When You're Stressed or Overwhelmed

PROMPT; Act as my thinking partner. Help me organize the scattered ideas below into a simple 3-point action plan I can begin today. Keep the tone supportive, clear, and low-pressure, with the aim of helping me take action without feeling overwhelmed. Please include: A brief summary of what the notes are truly about. A 3-point action plan written in plain, doable steps. One encouraging line that helps me feel confident starting. Messy Notes: [Paste your messy notes **here]** Goal: [Insert goal or outcome] Brand Voice (optional): [Insert brand voice]

This prompt alone can calm a busy mind within minutes.

Prompting doesn't just save time. It saves energy, reduces overwhelm, and gives you a reliable, repeatable path back into flow — even on days when your brain feels foggy or your schedule is overflowing.

You're not guessing anymore. You're guiding. And that's where real productivity begins.

Section Three

Progress You Can Feel — One Prompt at a Time

When you use the RIC framework, you don't just write a prompt. You create a repeatable system you can lean on anytime your mind feels full, your time is limited, or the blank page is staring at you with that familiar pressure. This is where the shift begins.

For most people, marketing feels complicated because they're constantly starting from scratch — juggling ideas, rewriting drafts, juggling tone, trying to "sound right."

But with one clear Role, one purposeful Input, and one thoughtful piece of Context, you've done what many professionals struggle to do: You've taken confusion and turned it into forward motion.

Three Plug-and-Go Prompts You Can Use Today

Here are three ready-made prompts you can copy, paste, and use immediately — each crafted for a high-impact marketing task. These are practical, realistic, and designed for busy marketers with limited time.

1. FOR A QUICK SOCIAL POST WHEN YOU'RE MENTALLY TIRED

PROMPT: Act as a social media copywriter. Turn the idea below into two short captions: one for Instagram and one for Facebook. Use a warm, relatable tone. Please include: A simple hook in the opening line. A friendly, easy-to-read rewrite of the idea. One relatable question at the end to encourage engagement. Keep both captions under 40 words. Idea: [Paste your idea] Brand Voice: [Insert brand voice] Audience: [Insert audience]

Why it works: It saves energy, guides engagement, and keeps branding consistent.

2. FOR A CLEAN, PROFESSIONAL EMAIL INTRODUCTION

PROMPT: Act as an email marketing specialist. Write a short introduction email for [describe your offer or product]. The tone should be confident, welcoming, and supportive, written for busy customers who want quick clarity. Please include: One clear benefit of the offer or product. A soft, friendly call to action that feels easy to follow. Keep the email under 120 words, simple, and easy to read. Audience: Busy customers Brand Voice: [Insert brand voice]

Why it works: It removes overthinking and gives you clarity within minutes.

3. For Getting out of a Mental Block or Foggy Headspace

PROMPT: Act as my clarity coach. Here are my scattered notes: [paste your messy thoughts]. Organize them into a simple 3-step plan I can follow today. Keep the tone calm, steady, and encouraging, with the goal of helping me regain focus without pressure. Please include: A brief reflection on what the scattered notes are really about. A 3-step plan written in clear, doable actions. One supportive line to help me feel grounded and ready to move forward. Goal: [Insert goal or outcome] Brand Voice (optional): [Insert brand voice]

Why it works: It clears mental clutter and instantly restores direction.

By completing this section, you've created something far more valuable than a single prompt. You've built: a new mindset around prompting — not as magic, but as structure, a three-part system (RIC) that turns messy ideas into clean, usable content, a reliable workflow you can return to whenever your time or energy is stretched thin, a practical method for clearing mental load quickly and professionally.

Closing Reflection

This chapter marked a shift from guessing to guiding—from forcing ideas to shaping them with intention. The RIC set up is more than a tool; it is a stabilising structure that keeps your message clear, even when your day is not. Clarity here is not theoretical. It is practical, repeatable, and usable under real conditions.

Forward into chapter Three

Chapter Three moves from mindset into structure. You will explore how to pair simple tools with AI, organise ideas and prompts, and build systems that reduce friction and save time. The focus is not on complexity, but on creating dependable workflows that turn scattered thoughts into clear action.

You do not need a large tech stack or expensive tools—only a few well-chosen pieces that work consistently. Take a pause here. The next chapter turns clarity into structure.

Chapter Three

Workflow Mastery for Busy Marketers

Section One

The more structure I give my AI process, the more freedom I create in my day. Once I established a workflow I could return to — even on tired days — the output became faster, cleaner, and far less stressful. I no longer lost time figuring out how to start. I knew the next step, and AI supported the gaps: drafting, formatting, refining, repurposing, and clarifying. The marketers who stay consistent are not the ones with the most, free time. They are the ones with a repeatable course of action that reduces decision fatigue, protects focus, and makes progress possible even when time is limited. This chapter walks you through building a system that is repeatable, forgiving, efficient, and designed for real-life marketing demands. Because when the method is simple, you use it. And when you use it, you make progress.

The Core 3 Workflow Layers

These three layers form the foundation of an AI-powered marketing system that works across industries, platforms, and schedules — especially a schedule that is already full. Think of them as **steps**, not rules. **Capture – Generate – Refine & Repurpose** When these layers work together, they eliminate the "start from scratch" problem and replace it with a workflow that keeps your marketing moving.

1) CAPTURE — Collect Ideas Before They Disappear

Every strong piece of content starts as a spark. And sparks are fragile. They show up during a client call, a conversation, while cooking dinner, during a scroll session, on a walk, or right before bed. If you do not capture them quickly, they disappear — and you end up back at the blank page, trying to force ideas that were already offered to you naturally.

That is why you need a **Prompt Inbox**: one simple place where you store anything that could become content later.

Where to capture. Choose one tool you already use consistently. Keep it simple. The tool does not matter – consistency does. Notes app on your phone, Google Doc; Notion or Trello; Canva doc; Voice memos' A physical notebook; Email drafts

What to capture, Capture anything that reflects real customer needs, real buyer psychology, or real moments worth teaching. client questions and objections; messy ideas and half sentences; hooks and headline drafts; frustrations you hear repeatedly; tips you explain often; stories, observations, or lessons learned; bullet points and rough outlines, screenshots of inspiration; questions you searched today; short wins and quick insights Content does not live inside a perfect schedule. It lives inside real life. Capturing is how you extract value from it. Plug-and-go prompt: turn scattered notes into usable content ideas

PROMPT: Act as my content organizer. Here are my scattered notes and ideas: [paste notes]. Turn them into five clear content ideas, each with: A simple, strong title A one-sentence description explaining the angle and purpose Keep the tone practical, clear, and immediately usable for a busy marketer or business owner. Goal: [Awareness / Education / Engagement / Promotion] Audience: [Insert audience] Brand Voice (optional): [Insert brand voice]

This instantly transforms clutter into usable building blocks – without you having to "think your way" into clarity.

2) GENERATE – Turn Ideas into First Drafts Without Overthinking

This is where ChatGPT becomes your strategic partner. Once an idea is captured, you do not start from scratch. You run it through a structured prompt using a simple professional framework:

The RIC Framework

Role, who AI is acting as (editor, strategist, copywriter, etc.)

Input; what you provide (idea, notes, excerpt, offer details)

Context; the audience, tone, platform, goal, constraints

This removes blank-page pressure and speeds up drafting because AI is no longer guessing what "good" means.

How to use this layer; Open ChatGPT; Copy one idea from your Prompt Inbox; Paste it into a RIC prompt; Receive a structured draft; Choose the best version and move forward

Example: messy input - structured output messy idea: "Small businesses overspend on ads because they don't target right."

RIC prompt to transform it:
REVISED PROMPT: Act as a brand storyteller for small business owners. Here is my rough idea: [paste idea]. Turn this into a short, 3-paragraph blog introduction written in a friendly, practical tone. Please include Paragraph 1: A relatable opening that connects with everyday challenges Paragraph **2:** A simple explanation of the main idea Paragraph 3: A practical lead-in that sets up the rest of the blog. Keep the writing warm, helpful, and easy to read. Audience: [Insert audience] Brand Voice (optional**):** [Insert brand voice]

What this produces. A polished first draft you can refine, repurpose, or publish — instead of fighting through mental fog.

3) REFINE & REPURPOSE; Multiply Output Without Doing More Work

This is the point where busy marketers save the most time. Once you have a solid AI-assisted draft, the goal is no longer "create more." The goal is to **reuse** what you already created across multiple channels. Think of your draft like a master asset. With the right prompts, it becomes social captions. LinkedIn posts. email teasers. newsletter paragraphs. carousel outlines. short video scripts. website copy. FAQs and objection reply. product descriptions and CTAs. Examples of refinement. make it shorter and clearer. rewrite for a different audience segment. improve tone consistency. reorganize structure. simplify complex explanations. remove repetition. strengthen hooks and CTAs. fact-check and personalize

Examples of repurposing

blog □ 7 social posts

email □ 3 reel scripts

video script □ newsletter intro

FAQ □ landing page copy

caption □ tweet thread

long email □ carousel outline

product description □ meta description

One idea becomes a content suite.

Plug-and-go repurposing prompt: turn one draft into five assets

PROMPT: Act as a content repurposing expert. Take the content below and create the following assets while keeping the same message, tone, and brand voice throughout: A short Instagram caption (friendly and scroll-friendly). A 150-word LinkedIn post (confident and professional). A 1-sentence email teaser (curiosity-driven). A headline (clear, strong, benefit-focused). A call-to-action (simple and aligned with the message) Content: [Paste content] Brand Voice: [Insert brand voice] Audience: [Insert audience]

This turns one draft into a full content suite — without rewriting from scratch. These workflow layers are not "extra steps." They are how you build sustainable marketing when time and energy fluctuate. They: simplify thinking, reduce overwhelm, eliminate the "start from scratch" loop, create consistent output, protect your time and attention, help you work even when you are tired, prevent you from losing good ideas, create momentum through repeatable wins When life is full, progress depends on systems — not stress.

Capture – Generate – Refine & Repurpose.
This is the master plan. Three steps that guide almost every marketing task without burning you out. With this in place, your marketing becomes stable, reliable, and sustainable — not because you are doing more, but because your system is doing more for you.

Section Two

Make It Work for Your Real Life

Real workloads include context switching, stakeholder changes, last-minute requests, platform shifts, client feedback loops, and personal responsibilities that cannot be "time-blocked away." So, the goal is not rigid structure. The goal is **a** repeatable operating rhythm that reduces decision fatigue and protects momentum. This section gives you a practical model you can use regardless of role, industry, or schedule: A weekly rhythm built on themes (not deadlines). multiple marketer **scenarios** to show how it flexes in real life, a clean weekly workflow table you can follow immediately

Principle

Rhythm Beats Schedule

A schedule is time-based: "I will write at 9:00am." A rhythm is outcome-based: "This week, I will brainstorm once, draft once, refine once, repurpose once, and schedule once." When your week gets disrupted (and it will), a rhythm prevents the common failure point: **You don't** lose the whole week — you simply resume at the next step. This makes your marketing system resilient. What This Rhythm Solves for Busy Marketers.

A practical weekly rhythm creates four operational advantages:

Reduces decision fatigue; You stop asking "What should I do today?" and start asking "Which step am I in?"

Prevents content chaos; You build from a small set of inputs (ideas + one anchor asset), not scattered tasks.

Protects consistency; Even a disrupted week still produces output because the workflow is modular.

Makes AI useful every week (not only when you're stuck) AI becomes a repeatable part of your process, not an occasional rescue tool.

To keep this rhythm realistic long-term, use these rules:

One anchor per week beats five scattered drafts: Choose one "main piece" (a post, a short email, a blog intro). Repurpose from it.

Batch when you can, not when you "should": If you get a clear window of energy, draft and repurpose in one session. Let the rhythm flex.

Protect the Friday finish: Scheduling + review prevents the next week from starting in panic mode.

Use AI to reduce decisions, not add steps: The purpose of AI here is speed + clarity. If a prompt creates more confusion, simplify it.

Once your rhythm is defined, the next step is choosing a small tool stack that makes execution effortless. In Section Three, I'll show you the core tools that support flow — not complexity — and how to use them as a practical system.

Section Three

Tools That Support Flow (Publisher-Grade Edition)

Modern marketers are surrounded by software promising speed, automation, and scale. In practice, too many tools fragment attention, increase maintenance overhead, and introduce unnecessary complexity. The result is often the opposite of efficiency: slower execution, more switching costs, and higher cognitive load. This section focuses on tool restraint. The objective is not to build a technology stack that looks impressive. The task is to build a reliable execution environment – one that supports clarity, consistency, and momentum even when time, energy, or focus is limited. What follows is a structured, professional overview of: the core tools that materially improve marketing workflow, how each tool fits into the Capture - Generate - Refine - Repurpose rhythm.

Before selecting any tool, apply this filter: Does this tool remove steps, or does it add them? A tool earns its place in your workflow only if it does at least one of the following: reduces decision-making; shortens time to first draft; improves clarity or consistency; prevents lost ideas; simplifies reuse and distribution If it does not clearly serve one of these outcomes, it is not essential. Regardless of industry or role, every sustainable marketing system requires tools that perform three distinct functions: Thinking + Drafting – Storage + Organisation – Distribution + Time Anchoring – Everything else is optional. The tools below are grouped by function—not by popularity—so you can clearly see where each one fits.

Tool Category 1: Thinking and Drafting

Primary Tool: ChatGPT

Function: Idea development, drafting, refinement, repurposing, planning, and clarity support. ChatGPT functions as the **intellectual engine** of your workflow. It accelerates thinking, reduces blank-page resistance, and provides structured output on demand. When paired with well-designed prompts, it replaces multiple manual steps: brainstorming – outlining – first drafts – rewrites – tone adjustments – format conversions – campaign mapping

What ChatGPT Does Best – Generates structured drafts quickly – Clarifies vague or scattered ideas – Produces multiple variations for selection –

Maintains consistent tone across assets.

Professional Use Case Examples

In-house marketer: Drafts campaign outlines, internal comms, and platform-specific variants without rewriting from scratch.

Agency marketer: Rapidly produces first drafts across multiple brands while maintaining tonal separation.

Solo consultant: Uses AI as a thinking partner to clarify positioning, offers, and thought-leadership content.

Small business owner: Creates usable content without spending hours thinking about what to say.

Recommended Usage Pattern

Use ChatGPT inside **short, focused sessions**, not open-ended exploration. For example: 20 minutes to generate drafts – 10 minutes to refine tone – 15 minutes to repurpose assets – Short sessions prevent over-editing and maintain momentum.

Tool Category 2: Storage and Organisation

Core Tools: Google Docs and Google Sheets

Function: Content storage, prompt libraries, draft management, repurposing logs, and lightweight planning. These tools are intentionally simple. Their strength is **low friction**. Complex tools are only effective if they are used consistently. Google Docs and Sheets succeed because they are: familiar – accessible on any device – fast to update – easy to share

How Google Docs Supports Workflow

Use Docs as your content home base.

Recommended documents: Prompt Library – Weekly Drafts – Repurposing Assets – Campaign Notes – Published Content Archive. This creates a searchable, editable library that grows over time.

How Google Sheets Supports Workflow

Use Sheets as your visibility and tracking layer.

Common uses: simple editorial calendar – content reuse tracking – subject line and hook library – posting status (Drafted / Scheduled / Live). This prevents duplicate work and makes your output visible at a glance.

Why These Tools Matter: They eliminate two major workflow failures: losing good ideas – forgetting what you've already created – Organisation does not need to be complex to be effective. It needs to be **consistent**.

Tool Category 3: Distribution and Time Anchoring

Primary Tools: Calendar + Scheduling Platform

Function: Anchoring content into real time. No matter how efficient your drafting process is, marketing only works when content is **published**. A calendar—digital or paper—connects your workflow to reality. It ensures ideas move from concept to execution.

Scheduling Tools (Optional but Useful) Examples include: – Meta Business Suite – Buffer – Hootsuite – Later- Native platform schedulers – Use one tool only. Multiple schedulers increase friction. – Scheduling should be a final step, not a daily task.

Minimum Viable Tool Stack

If your time is limited and you want the simplest possible system, this is sufficient: The Minimum Stack

ChatGPT – thinking, drafting, repurposing

Google Docs – prompt library and drafts

One calendar – anchoring work into time

That is enough to: plan content – write drafts – reuse ideas – stay consistent - avoid burnout - Everything else is optional. Weekly content creation in under 2 hours - Reduced mental load - Consistent messaging = Lower resistance on low-energy days - Sustainable output over time - This stack is deliberately small, so it remains usable even during busy or disrupted weeks.

Optional Expansion Tools (Only If Needed)

These tools are not required but can be helpful when your workflow becomes more complex.

Notion or Trello - Use only if: You manage multiple projects, clients, or campaigns. - Best for: - visual pipelines - campaign tracking - long-term planning- Keep boards simple. Avoid over-engineering.

Canva - Use if: You create visual content regularly. - Best for: - carousels - social graphics - brand consistency - Use templates to reduce effort.

Email Platforms (Mailchimp, Convert Kit, etc.) Use if: Email is a primary channel. - Best for: - sequencing - segmentation - automation - Draft content in Docs first, then transfer.

Tool Discipline: What to Avoid - Avoid: - adopting tools without a clear purpose - switching platforms frequently - building systems you must "maintain" - over-customising dashboards - using tools as procrastination A system that requires constant tuning will eventually be abandoned.

The Strategic Advantage of a Lean Tool Stack

When your tools are stable: your workflow becomes predictable - your output becomes consistent - your confidence increases - your energy is protected - You stop reacting and start operating.

This is what allows marketers to show up reliably without burnout.

Closing Reflection

When you choose a small set of tools and use them well, your workflow becomes dependable. Friction decreases, cognitive load lightens, and marketing begins to feel calmer and more deliberate. This creates space for strategic thinking rather than constant execution pressure.

The goal is not to do more. It is to make what you already do easier to repeat. This is where rhythm turns into long-term results.

Forward into chapter Four

Chapter Four focuses on sustained momentum—how to maintain consistency without burnout as demands increase and energy fluctuates. You will learn to focus on what matters, break content into manageable steps, and use small, repeatable actions to build flow. We will explore micro-planning, batching, and AI-supported workflows that accelerate execution without compromising your voice.

Chapter Four

Building Your Personal Prompt Library

Section One

Before I understood how powerful a plug-and-play prompt could be, my prompting process was... well, let's just say it was spirited but inefficient. A prompt library isn't just helpful; it's essential. It saves time, energy, and frustration, and it keeps your creativity flowing instead of getting stuck on the starting line. Once I experienced the freedom of a plug-and-play prompt, everything changed. My confidence grew. My writing flowed faster. I stopped dreading the blank page. And for the first time, I acknowledged I didn't have to rely on willpower alone — I could rely on structure. This is exactly why I want you, dear reader, to have your own prompt library. Not to replace your creativity, but to support it. Not to restrict you, but to free you. Because the right engineered prompt system doesn't just save time - it gives you back a sense of ease, permission, and energy... the kind that lifts your whole day. That lesson came wrapped in humor and a little nudge from Sage — and it opened the door to a whole new way of creating

Plug-and-play prompts for marketers who need results, not guesswork

If there is one thing that slows marketers down more than anything else, it's this: The blank page. Not because you don't have ideas. Not because you're not creative. But because you've already written versions of the same content dozens — if not hundreds — of times before.

1). Welcome emails. 2) Social posts. 3) Monthly promotions. 4) Customer replies. 5) Product descriptions. 6) Profile bios.

This is where your Prompt Library becomes one of the most powerful tools in your marketing blueprint — not because it stores information, but because it stores clarity. A Prompt Library isn't a database. It isn't a filing system. It isn't one more thing to "manage." It's your shortcut. Your spark. Your momentum. Your way out of the mental fog. It's the bridge between "I don't know where to start" and "I can get this done in ten minutes."

What a Prompt Library Really Gives You

1. Speed — the kind you can feel instantly. When time is tight, you don't have the luxury of warming up or waiting for inspiration. A good prompt library removes the hesitation and hands you the starting line. Instead of thinking, "*What should I write?*" you think, "*Which prompt do I want to use?*" That small shift saves hours over a month.

2. Confidence — because you know what works. A blank page can make even experienced marketers doubt themselves. A prompt library removes the insecurity. Every prompt in your library is something you've tested, refined, and proven. It's your voice. Your tone. Your rhythm. Your marketing heartbeat captured in words. This is especially valuable for people juggling many roles — parents, small business owners, employees, creators — because confidence saves mental energy. And mental energy is gold.

3. Consistency — without burning out. You don't have time to reinvent yourself with every post or email. A prompt library gives you: consistent tone, consistent structure, consistent output, consistent branding, even when your week is chaotic, your content stays stable.

4. Creativity — because structure creates mental space. Constraints are often misunderstood as limitations. In practice, for marketers operating under real-world pressure, clear structure reduces friction and allows creative thinking to surface more easily.

5. Structure —creates freedom. When you have a prompt that handles structure, your brain is free to add personality, story, and meaning — without the pressure of doing everything at once. Your Prompt Library becomes the supportive foundation that carries your creativity, instead of draining it.

Why This Matters for You. Some days your energy is stretched. And your process needs to support you, not weigh you down. That's why a prompt library matters. It gives you a reliable, repeatable way to move forward – even on the days when your mind feels scattered, your focus is limited, or you simply don't have the bandwidth to start from scratch. A strong engineered prompt system transforms your marketing life from: Start from scratch - Start from clarity – From: Stress – Structure – From: Overwhelm - Flow

What You Can Use a Engineered Prompt Library For

Here are just a few of the everyday tasks a prompt library makes easier: Welcome emails, Weekly newsletters Instagram captions, Facebook posts, LinkedIn articles, Product descriptions, Monthly promotions, Service explanations, Customer replies, lead magnet descriptions, Bio rewrites, Testimonial requests, Follow-up emails, Content repurposing, Video scripts. Any task you repeat more than twice should have a dedicated prompt in your library. Because your time matters. Your energy matters. And your voice deserves a system that brings out its best.

The Real Shift: Once your Engineered Prompt Library is built, marketing stops feeling like guesswork. You stop facing blank pages. You stop wasting time trying to remember "how" to start. Instead, you open your library... pick a prompt... paste it into ChatGPT... and you're already halfway done. It's one of the most powerful forms of self-support you can give yourself — and one of the simplest ways to unlock clarity and momentum in your weekly marketing.

Section Two

The 5 Prompt Types Every Marketer Needs

Over time, I found that there are only five types of prompts I use every single week, whether I'm creating content for a business, planning a campaign, or promoting a book. Once you understand how these five work, you'll be able to produce content faster, reduce friction in your day, and keep your message fresh without burning yourself out. These are the prompt types they are practical enough for beginners, powerful enough for professionals, and flexible enough for life that doesn't always go to plan.

1. Social Caption Prompts: Social media captions don't need to be complicated. In fact, short and clear usually wins. But writing captions on demand? That's the part that drains your energy. A good social caption prompt helps you skip the struggle and go straight to something usable.

PROMPT: Act as a social media manager for a brand that helps [target audience]. Write a caption that introduces [topic/product] using: A casual, relatable hook in the opening line One clear benefit that matters to this audience A relatable call-to-action that feels natural and encouraging Keep the tone friendly, helpful, and easy to read. Topic/Product: [Insert topic or product] Brand Voice: [Insert brand voice] Audience: [Insert audience]

Why This Helps: Removes guesswork. Delivers captions that sound human. Keeps your tone warm and conversational. Helps you stay consistent across platforms

PROMPT A: Act as a social media manager for an author who helps busy marketers. Write an Instagram caption introducing my book Prompt Engineering for Busy Marketers. Include: A relatable hook that speaks to the everyday challenges marketers face One clear, results-focused benefit of the book A simple, encouraging CTA inviting readers to learn more Keep the tone friendly, confident, and supportive, and ensure the message is easy to read for time-poor professionals. Brand Voice: [Insert brand voice] Audience: Busy marketers and business owners

PROMPT B: Act as a brand storyteller. Write a Facebook post explaining why my book Prompt Engineering for Busy Marketers is ideal for marketers juggling real-life responsibilities. The tone must be warm, practical, and encouraging, speaking directly to the realities faced by time-poor professionals. Please include: A relatable opening that acknowledges the challenge of balancing work, life, and marketing tasks A simple explanation of how the book helps — emphasizing clarity, time-saving, or confidence One encouraging line that makes the reader feel understood and supported A gentle call-to-action inviting them to explore or learn more about the book Brand Voice: [Insert brand voice] Audience: Marketers navigating busy schedules and multiple responsibilities

2. Email Prompts: Email is where trust is built — but it can also be the hardest channel to write for because it requires clarity, flow, and a personal touch. A strong email prompt gives you structure so you're not staring at a blinking cursor.

PROMPT: Act as an email copywriter for a values-based business. Write a 3-paragraph email that shares [topic/value/tip] in a helpful, friendly tone and includes one simple call to action. Please include Paragraph 1: A warm, relatable opening that connects the topic to everyday experience Paragraph 2: A clear explanation or practical insight based on the topic/value/tip Paragraph 3: A supportive closing with one simple CTA that feels natural and encouraging Keep the email easy to read, supportive, and aligned with values-driven communication. Brand Voice: [Insert brand voice] Audience: [Insert audience]

Why This Helps: Creates natural, easy-to-read emails. Reduces overthinking. Keeps your tone authentic and grounded. Saves time drafting

PROMPT A: Act as an email copywriter. Write a 3-paragraph email announcing my new book Prompt Engineering for Busy Marketers. Please include: Paragraph 1: A clear, engaging introduction announcing the book and who it's for Paragraph 2: One key problem the book solves for busy marketers and why it matters Paragraph 3: A friendly invitation to read the first chapter, framed as an easy next step Keep the tone confident, warm, and helpful, and ensure the message is easy for busy readers to understand quickly. Brand Voice: [Insert brand voice] Audience: Marketers juggling real-life responsibilities

PROMPT B: Act as a marketing coach. Write a warm, supportive email to my subscribers explaining how this book can help them manage content creation with less stress. Please include: A relatable opening that acknowledges the challenges of keeping up with content A simple explanation of how the book makes content creation easier, clearer, or more manageable One encouraging line that reinforces confidence and reduces pressure A gentle call-to-action inviting them to explore or learn more about the book Keep the tone warm, human, and reassuring, and make the email easy for busy readers to digest. Book Title: Prompt Engineering for Busy Marketers Brand Voice: [Insert brand voice] Audience: Time-poor marketers and business owners

3. Offer or Product Description Prompts: Whether you're selling a book, a service, a course, or digital download, you need a description that is clear, quick, and compelling.

PROMPT: Act as a conversion copywriter. Write a short product description for [offer/service] that clearly states: What it is Who it's for the main benefit Keep the description **under** 80 words, concise, compelling, and easy for busy readers to understand. Audience: [Insert audience] Brand Voice: [Insert brand voice]

Why This Helps: Keeps details tight and focused. Makes your offer easy to understand. Perfect for Amazon, websites, landing pages, and social media

PROMPT A: Act as a product marketer. Write an 80-word book description for Prompt Engineering for Busy Marketers that clearly highlights: The problem the book solves Who will benefit most from it The description should be concise, compelling, and easy for busy readers to understand. Focus on clarity, value, and real-world usefulness. Brand Voice: [Insert brand voice] Audience: Marketers and business owners managing heavy workloads

PROMPT B: Act as a landing-page copywriter. Write a short, persuasive blurb explaining why time-poor marketers need this book in today's fast-paced digital world. Please include: One clear pain point marketers face in their daily workflow A simple, compelling benefit the book delivers A forward-looking statement showing why this matters now Keep the blurb concise, high-impact, and easy to read Book Title: Prompt Engineering for Busy Marketers Brand Voice: [Insert brand voice] Audience: Time-poor marketers navigating rapid digital change

4. Content Repurposing Prompts: This is where your workload gets lighter. Instead of creating NEW content, you turn one piece into many — without losing your tone or message. Multiples your effort. Saves hours. Helps you stay visible across platforms. Keeps your voice consistent

PROMPT A: Act as a content repurposing expert. Turn the book excerpt below into a LinkedIn post that highlights one key insight and positions the author as a helpful, practical voice in the marketing space. Please include: A strong opening line that captures attention in the LinkedIn feed A concise rewrite of the excerpt that keeps the core message intact One standout insight that would matter to busy marketers or industry professionals A short closing line that encourages reflection or conversation Keep the tone confident, clear, and professional, suitable for LinkedIn. Book Excerpt: [Paste excerpt] Brand Voice: [Insert brand voice] Audience: Marketing professionals and business owners

PROMPT B. Act as a social strategist. Turn the introduction of my book into five short social posts designed to inspire overwhelmed marketers to believe they can learn AI. Please include: Five standalone posts, each under 40 words An encouraging, empowering tone that removes pressure and builds confidence One clear idea per post, drawn directly from the introduction's message Simple language suitable for time-poor marketers scrolling quickly Optional: Suggest one hashtag per post to support reach Keep the tone warm, practical, and motivational, and ensure each post reinforces the message that AI is learnable, accessible, and doable. Book Introduction: [Paste introduction] Brand Voice: [Insert brand voice] Audience: Overwhelmed marketers who want clarity and confidence with AI

5. Voice Refinement Prompts: Sometimes the idea is strong, but the tone isn't quite right. Maybe it's too formal, too stiff, or too generic. A voice refincment engineered prompt keeps your message intact while adjusting the tone to something more you.

PROMPT: Rewrite the content below using a tone that is [warm, confident, direct, casual, supportive, etc.]. The audience is [describe your audience]. Keep the original structure and flow, but adjust the voice, word choice, and emotional tone to match the new style. Please include: A polished rewritten version that maintains the same format as the original Consistent messaging, with no added or removed ideas Tone alignment that fits the specified audience and desired style Original Content: [Paste content here] Tone Required: [Insert tone] Audience: [Insert audience] Brand Voice (optional): [Insert brand voice]

Why This Helps: Ensures your content always sounds like you. Saves time rewriting. Helps polish your message without losing authenticity

PROMPT A: Rewrite the paragraph below from my book description in a tone that is warm, steady, and confident. My audience is time-poor marketers who need encouragement, not pressure. Please ensure: The meaning stays the same, but the emotional tone softens and strengthens at the same time The wording feels calm, reassuring, and grounded, while still conveying authority The message is clear, supportive, and easy to understand quickly The rewrite avoids urgency or overwhelm, and instead promotes clarity, confidence, and ease Original Paragraph: [Paste the paragraph here] Brand Voice (optional): [Insert brand voice]

PROMPT B: Rewrite the caption below so it sounds more like an author speaking directly to a busy reader. Please ensure: The tone feels personal, grounded, and human The message speaks directly to the reader's real-life pace and challenges The voice reflects an author's supportive guidance, not a brand announcement The rewrite remains clear, concise, and reader-friendly Original Caption: [Paste caption here] Audience: Busy readers looking for clarity and encouragement Brand Voice (optional): [Insert brand voice]

Why These 5 Prompt Types Matter These five engineered prompt types are the backbone of an efficient, sustainable marketing system. They reduce mental load, save valuable time, and remove the daily decision fatigue that slows momentum. For marketers, these prompts offer a practical way to stay visible, stay confident, and stay in motion. You don't need hundreds of prompts. You simply need the right ones – the type of prompts that move your message forward. These five prompts will transform your workflow, shape your marketing rhythm, and strengthen your confidence. And as you grow, your Prompt Library will grow with you.

Section Three

How to Build and Maintain Your Prompt Library (Without Feeling Overwhelmed

You've gathered the engineered prompts, you understand the categories, and you've seen how they work. Now it's time to take the most important step: Put your Prompt Library to work in your real, everyday life. Because a prompt library isn't just a list of clever words. It's your personal productivity toolkit. A fast-access folder of ideas, templates, and writing momentum – ready whenever you are. That's the power of this prompt system. It brings calm where there used to be pressure. It brings clarity where there used to be scattered thoughts. It brings momentum where you used to lose minutes – sometimes hours – overthinking. Let me show you how to put it into practice.

1. Google Doc – The Master Library. This is my home base. One long, clean, reliable document divided into bolded categories: Social. Email. Offers & Promotions. Repurposing. Voice & Tone. Book Marketing/ Clarity & Strategy Under each category, I paste: my favorite prompts, my improved versions, small notes about what worked best, It's simple. It's searchable. And it's always there when I need it – on my laptop, phone, or tablet.

2. Notion or Trello – For Visual Compose: If you like drag-and-drop clarity, Notion or Trello will feel like a breath of fresh air. **You can create:** A card for each prompt. A space inside each card for your best outputs. A column for "To Try," "In Use," and "Favorites". A color coding for different content types This helps you see your creative workflow at a glance, especially if you handle multiple clients, platforms, or projects.

3. Printed Cheat Sheet – The Old-School Secret Weapon: Yes, paper still works – and sometimes it works best. I keep a one-page cheat sheet near my desk with the engineered prompts I use every week. Those quick wins matter, especially on days when my mind feels tired or pulled in ten directions. A printed prompt makes it easy to act – without switching tabs or losing focus. This is especially helpful when you only have five minutes and need a caption, headline, or email intro *now*.

What to Add Over Time (For a Library That Grows with You)

Your Prompt Library isn't something you build once and forget. It evolves – just like your business, your marketing, and your confidence. Here's what you'll naturally add:

1. Prompts that helped break writer's block: Whenever a prompt gets you unstuck, save it. This becomes your emergency kit.

2. Revised versions with better phrasing: Over time, you'll tweak prompts until they "click." Keep those improved versions – they are gold.

3. High-performing prompts: Notice a caption that got strong engagement? A newsletter with high open rates? A product description that improved conversions? Save the exact prompt that created it.

4. Prompts specifically for book marketing: Your book is part of your brand. Keep prompts like: "Write a new author bio using my story but in a shorter format." "Turn this chapter excerpt into a social teaser." "Write a newsletter announcing a book giveaway." The more you save, the more you'll discern that you're building a toolkit tailored to your life, your brand, and *your* message.

Closing Reflection

This chapter introduced a subtle but meaningful shift: consistency is not built through discipline alone, but through design. You explored ways to create flow by working with your mind rather than against it—using batching, mini-targets, prompts, and momentum-building techniques to make content creation feel lighter and more achievable.

These approaches support confidence and stability without relying on intensity. This is how sustainable habits begin to take shape. That's what keeps your marketing steady without burning you out.

Forward into Chapter Five

Chapter Five focuses on alignment—matching the right tool to the right task. You will learn how to identify repeat work, reduce time and energy drains, simplify your toolkit, and use AI intentionally to turn complex tasks into clear workflows. The emphasis is not on adding more tools, but on reclaiming time through clarity and effective delegation. it's all about empowerment, ease, and reclaiming your time.

Chapter Five

Campaigns, Not Chaos

Section One

When I first began experimenting with AI, I began to recognize a pattern, I noticed, it wasn't my prompting that was powerful... it was the momentum created when I linked prompts together. One prompt □ became the revised and restructured input for the next□ which shaped the next □ which built the next... And before I knew it, I was learning how to engineer prompts, I wasn't just creating captions or headlines anymore. I was building entire marketing campaigns — step by step, in a way that felt clear, calm, and achievable.

This chapter is about exactly that: Prompt Engineered Workflows. Not standalone prompts... but smart, simple prompt chains that work together like building blocks. This isn't theory. And once you start thinking this way, you'll never go back to chaos again. A single prompt can generate a single asset. A grand design generates a system: one message, many deliverables, all aligned.

That awareness matters because marketing rarely fails due to lack of ideas. It fails when the output is disconnected — content that doesn't match, messaging that drifts, campaigns that start strong and then stall because there's no sequence holding them together.

Prompt chains solve that by creating a repeatable path. You define the message once, then reuse it as structured input across social, email, landing pages, follow-ups, and FAQs. Instead of reinventing the campaign every time, you're simply moving through a consistent set of steps. Most importantly, you will walk away with prompt chains you can use for: book launches and promotions; small digital products and lead magnets; services and seasonal campaigns; collaborations and affiliate partnerships;

ongoing "stay visible" marketing without random posting.

These chains are designed for real workloads. You'll see exactly how to move from spark □ structure □ execution with clarity and control.

The Rules of Prompt Chains (So Your Outputs Stay Clean and Consistent)

Prompt chains work best when you treat AI like a production assistant — not a mind-reader. The quality of the chain depends on how stable your inputs are.

Here are the key rules that keep chains producing consistent, usable marketing assets:

Rule 1: Lock the message before you multiply the content. If you start re-purposing before your core message is clear, you end up creating ten different versions of confusion. Always start with the core statement first: who it's for, what it solves, and what outcome it creates.

Rule 2: Reuse the same "message map" as your source of truth. Think of your message map as campaign DNA. Every post, email, and landing page section should be generated from the same core inputs. This is how you keep tone, terminology, and benefits consistent across channels.

Rule 3: Use constraints to protect quality. Professional outputs require boundaries: word count, structure, tone, CTA style, platform format, and audience context. Constraints don't reduce creativity — they reduce rework

Rule 4: Don't chain noise. Chain clean inputs. If an output is messy, unclear, or off tone, refine it before feeding it into the next step. Prompt chains are compounding systems: they amplify clarity, but they also amplify chaos if you allow it into the pipeline.

Rule 5; Save the best chain as a reusable template. Once a chain performs well, it becomes an asset. Save it, label it, and reuse it. This is how you build a marketing system that gets stronger over time — rather than starting from scratch each week.

When you follow these rules, prompt stacking becomes predictable. And predictability is what creates sustainable marketing. Most busy marketers don't struggle with ideas — they struggle with structure. You get a great product idea, but: what do you write first? which platform should you start with? what should the message be? how do you keep everything consistent? how do you make sure every piece connects? This is where prompt stacking becomes your secret weapon. With one spark of an idea, you can guide AI to create an entire funnel: your core message, social posts, email sequence, landing page outline, product description, follow-up message. promotional angles, captions and teasers, FAQ section and more. All created through structured steps — not stress. Once you master this approach, you don't just "use AI." You *direct* it. You become the strategist, and AI becomes the assistant building your assets.

Imagine you start with nothing more than a rough idea:

PROMPT: I want to launch a simple digital checklist to help busy marketers plan their weekly content.

With prompt structures, that tiny idea becomes a full campaign in minutes. Here's what it looks like when one output becomes the next input:

Prompt 1: Generate a Big Idea: Prompt AI to clarify your core message.

Prompt 2: Turn the Message into a Campaign Map*:* AI outlines the themes, hooks, angles, and audience benefits.

Prompt 3: Build Social Posts: Using the message map as input.

Prompt 4: Build an Email Sequence: Using the same message map as the foundation.

Prompt 5: Build the Landing Page Outline: Using the email and social content as context.

Prompt 6: Write a Follow-Up Message: Using the final output to ensure consistency.

Campaign Assets You Can Generate From One Message

Once you have a clear core message, you can generate a complete campaign pack quickly. Depending on your offer, a single workflow can produce: 3–5 social captions (educational, relatable, promotional); a 3-post teaser sequence (awareness □ curiosity □ value); a launch-day email + reminder email; a landing page outline (hook, problem, benefits, CTA); 5–7 FAQ entries based on objections; 3 DM replies for "interested / send it / tell me more"; 5 headline variations and 3 subhead line options; one "what this is / who it's for" product description

When these are generated from the same message map, the output doesn't just save time — it builds confidence. You know every piece connects, because every piece was built from the same foundation. Each step builds on the last.

This isn't complicated. This isn't overwhelming. This is a system that gives you: clarity, cohesion, speed, momentum, consistency and confidence. All while keeping your voice at the heart of every step.

The Real Transformation: Prompt stacking turns you from: "I don't know where to start." "This feels chaotic." "Every piece sounds different." "I'm writing from scratch again." Into: "I know the next step." "My message is consistent." "I can create full campaigns quickly." "My content finally has a structure." You're no longer jumping between random pieces of content. You're creating a coherent campaign with a calm, steady flow — one prompt at a time. It doesn't matter if you're promoting: a book, a digital product, a service, a local business, a workshop, or a simple idea.

Prompt workflows help you move from spark □ strategy □ execution with ease. This is how you take your marketing from scattered to streamlined. From chaos to clarity. From "too much work" to "I can do this." And once you experience this, you will never go back.

Section Two

The 6 Essential Prompt Chains Every Marketer Should Use

Instead of writing content piece by piece, you build a step-by-step flow where every prompt supports the next. This creates clarity, consistency, and momentum — even when life is busy and your brain feels scattered. Below are the six prompt chains I rely on when I want to launch something small but meaningful...or promote a service... or highlight a new offer... or simply stay visible with value-led messaging.

These chains work beautifully for: digital products, books, workshops, services, local businesses, seasonal offers, or any message you want to spread consistently

CHAIN 1: Define - Social – Email:

This gives you a short, powerful statement you can use everywhere.

Use this chain when: [describe scenario]; Ideal outcome: [what it produces]; Quality standard: voice consistency; clear CTA style; alignment to "Campaigns, Not Chaos"; minimal edits required

Step 1: Define the Core Message

Before you write anything else, you need clarity. This step gives you the message that holds the entire campaign together.

PROMPT: . Act as a brand messaging strategist. I'm launching [product/service/book]. Help me define the core message in 2–3 sentences. Include: who it's for, what problem it solves, and the result it helps them achieve. Tone: clear, confident, and genuine.

Step 2: Generate Social Content Variations.

Now take your core message and spread it across platforms — without reinventing anything.

PROMPT: Act as a content repurposing expert. Take this core message and turn it into a short Instagram caption, a casual, helpful LinkedIn post, a 5-slide carousel outline Keep everything aligned with the original tone.

Step 3: Create a Warm Follow-Up Email or DM.

This step is where most marketers fall off — but follow-up is where conversations become conversions.

PROMPT: Act as a relationship-based marketer. Write a friendly, soft follow-up email/DM for someone who showed interest but didn't buy. Keep it warm, pressure-free, and helpful.

CHAIN 2: Idea - Headline - Landing Page:

This chain turns a simple idea into a full marketing asset fast.

Use this chain when: [describe scenario]; **Ideal outcome:** [what it produces]; **Quality standard:** voice consistency; clear CTA style; alignment to "Campaigns, Not Chaos"; minimal edits required

Step 1: Idea to Hook

PROMPT: Act as a creative strategist. Turn this idea into 5 compelling hooks: [paste idea]. Tone: clear, bold, and curiosity driven.

Step 2: Hook to Headline

PROMPT: Act as a conversion copywriter. Turn this hook into 3 landing page headlines.

Step 3: Headline to Landing Page Outline

PROMPT: Act as a landing page architect. Using this headline, create a simple landing page outline: Hook. Problem. Benefit. Offer. CTA

CHAIN 3: Blog - Email - Social:

This chain helps you maximize the value of every long-form piece you create

Use this chain when: [describe scenario]; Ideal outcome: [what it produces]; Quality standard: voice consistency; clear CTA style; alignment to "Campaigns, Not Chaos"; minimal edits required

Step 1: Blog to Email

PROMPT: Act as a newsletter writer. Turn this blog into a short, friendly email that highlights the key takeaway.

Step 2: Blog to 3 Social Posts

PROMPT: Act as a repurposing expert. Turn this blog into 1 educational post. 1 inspirational post. 1 practical tip post

Step 3: Blog to a Reel or TikTok Script

PROMPT: Summarize this blog into a 20–30 second video script with a hook, middle, and CTA.

CHAIN 4: Testimonial ⊠ Story ⊠ Sales Angle:

If you have testimonials, this chain turns them into content you can use everywhere — without sounding "salesy.

Use this chain when: [describe scenario]; Ideal outcome: [what it produces]; Quality standard: voice consistency; clear CTA style; alignment to "Campaigns, Not Chaos"; minimal edits required

Step 1: Testimonial to Story Post

PROMPT: Turn this testimonial into a relatable story-style post that highlights the transformation.

Step 2: Story to Email

PROMPT: Turn this story into a 2-paragraph email with a soft CTA.

Step 3: Story to Sales Angle

PROMPT: Turn this story into 3 sales angles for this offer.

CHAIN 5: Outline - Script -Carousel:

Perfect for turning educational content into multiple formats.

Use this chain when: [describe scenario]; Ideal outcome: [what it produces]; Quality standard: voice consistency; clear CTA style; alignment to "Campaigns, Not Chaos"; minimal edits required

Step 1: Outline

PROMPT: Act as a content designer. Create a simple outline for [topic].

Step 2: Script

PROMPT: Turn this outline into a short teaching script.

Step 3: Carousel

PROMPT: Turn this outline into a 5-slide carousel.

CHAIN 6: Offer – Objections – FAQ:

This chain builds trust by addressing concerns before they arise.

Use this chain when: [describe scenario]; Ideal outcome: [what it produces]; Quality standard: voice consistency; clear CTA style; alignment to "Campaigns, Not Chaos"; minimal edits required

Step 1: Objection Finder

PROMPT: Act as a customer insight analyst. List the top objections people might have about [offer/book/service].

Step 2: Objection Answers

Prompt: Write supportive, clear responses to each objection.

Step 3: Build an FAQ Section

PROMPT: Turn these objections and answers into a friendly FAQ for my website or sales page.

Why These 6 Chains Work. They give you: a proven structure, a consistent voice, a predictable workflow, faster creation, stronger confidence, and a clear path from idea □ execution. No chaos. No overwhelm. No guesswork. Just clarity, flow, and content that feels like you – created faster and with less effort. To make prompt chains sustainable, you need a simple way to store and reuse them. Keep it basic: Create one folder called "Campaign Chains."; Inside it, create subfolders: Launch, Lead Magnet, Collaboration, Evergreen, Repurposing. Save your "message maps" separately from your outputs.

Your message map is the master file. Outputs come and go. The map is what you reuse. Name chains by purpose, not platform; Example: "Lead Magnet: Define □ Social □ DM Replies" is more reusable than "Instagram Chain."Keep a 'Best Performing' file; When something performs well (comments, opens, clicks), save that version. You're building a library that improves over time.

Section Three

Build a Fully AI-Assisted Campaign (Step by Step)

This is the moment your personal system transforms into a marketing engine – one that works with you, supports you, and keeps your message moving even when your week doesn't go to plan. Once you've practiced the 3-step campaign workflow (Define □ Social □ Follow-Up), you'll notice something powerful: You can use the exact same structure for almost any marketing goal — without more tools, more stress, or more complexity. This section shows you how to build a full, AI-assisted campaign from start to finish. Everything here is flexible, repeatable, and designed for real-life marketers who want clarity, not chaos.

Let's walk through three of the most common campaign types

1. Launching a New Offer: This workflow works beautifully for digital products, workshops, services — and yes, book launches.

Step 1 – Define the Offer's Outcome + Ideal Client. Before anything else, get crystal clear on what you're offering, who it helps, and what outcome it gives.

PROMPT: Act as a brand messaging strategist. Help me define the core message for my new [offer/book/service]. Write a short, polished message that includes Who it's for What problem it solves the transformation it provides Use a tone that is confident, clear, and supportive, speaking directly to busy professionals who want organized, effective marketing — not chaos. Audience: [Insert audience] Brand Voice: [Insert brand voice]

Step 2 – Turn the Message Into a 3-Part Teaser Sequence: Think of this as building anticipation before the official launch.

PROMPT: Act as a launch content creator. Turn the core message below into a three-post teaser sequence designed to build: Awareness, Curiosity, Value Keep the tone warm, confident, and encouraging, speaking directly to busy professionals who want organized, effective marketing without overwhelm. For each of the three posts, include: A clear purpose (Awareness / Curiosity / Value) A short post (30–50 words) A gentle lead-in that supports the theme "Campaigns, Not Chaos" A simple, optional next step Core Message: [Paste your core message here] Audience: [Insert audience] Brand Voice: [Insert brand voice]

Step 3 – Draft a Warm Launch-Day Email + Reminder Launch emails don't need to be loud. They need to be warm, clear, and human.

PROMPT: Act as an email copywriter. Write two emails for launch: Email 1: Launch-Day Email A warm, friendly email introducing this offer. Keep it simple and clear. Include: Why it matters Who it helps One key benefit A link to purchase Tone: warm, calm, encouraging — aligned with "Campaigns, Not Chaos." Email 2: Next-Day Reminder A short, friendly reminder email (under 80 words) that: Reinforces one core benefit Acknowledges busy schedules Gently invites them to take a look again Offer: [Insert offer/book/product] Audience: [Insert audience] Brand Voice: [Insert brand voice]

2. Promoting a Lead Magnet: This is how you grow your audience without pressure or pushiness.

Step 1 – Define the Value in Plain Terms. Skip the jargon. Go straight to the benefit.

PROMPT: Act as a clarity coach. Explain the value of this lead magnet in simple, everyday language. Clearly state: What problem it solves What someone will walk away with Keep the tone warm, calm, and supportive, aligned with the theme "Campaigns, Not Chaos." Focus on helping busy people understand how this resource makes their marketing easier and less overwhelming. Lead Magnet Description: [Insert details] Audience: [Insert audience] Brand Voice: [Insert brand voice].

Step 2 – Create a Carousel or Reel Script Introducing It: People engage more when content is visual and fast-moving.

PROMPT: Act as a social media strategist. Create either: a 5-slide carousel, or a 20-second reel script to introduce this lead magnet. Include: A strong hook that grabs attention fast The value of the lead magnet in simple, benefit-focused language Who it's for (speak directly to busy, overwhelmed marketers) What they'll learn or walk away with A clear call to action aligned with "Campaigns, Not Chaos" Keep the tone warm, confident, supportive, and easy for time-poor audiences to absorb quickly. Lead Magnet: [Insert description] Audience: [Insert audience] Brand Voice: [Insert brand voice] Format Preference: [Carousel or Reel]

Step 3 – Draft a Reply or DM (replying privately to someone) Message: This is essential for engagement.

PROMPT: Act as a relationship-based marketer. Write a friendly reply message for someone who comments "send it" or "interested." The tone must be warm, simple, and supportive, aligned with the theme "Campaigns, Not Chaos." Please include: A brief thank-you or acknowledgment A simple line explaining what they'll receive or what it's about One supportive line that reinforces ease, clarity, or "no-pressure" A clear, friendly next step (link, DM, or instruction) Keep the message short, human, and easy to read. Offer/Lead Magnet/Product: [Insert details] Audience: Busy, real-life marketers Brand Voice: [Insert brand voice]

3. Creating a Collaboration or Affiliate Campaign: Collaborations help you reach entirely new audiences – but your message still needs structure.

Step 1 – Clarify the Partner's Value Get clear on why the partnership matters.

PROMPT: Act as a collaboration strategist. Explain the value this partner brings to my audience in 2–3 warm, conversational sentences. Please include Who the partner is or what they offer Why their expertise or service matters to busy, real-life marketers How the partnership supports clarity, simplicity, or momentum, aligned with "Campaigns, Not Chaos" Keep the tone friendly, genuine, and easy to understand. Partner Details: [Insert partner info] Audience: Busy marketers seeking guidance and simplicity Brand Voice: [Insert brand voice]

Step 2 – Draft Introductory Posts That Educate + Excite: These posts help your audience understand why the collaboration is meaningful.

PROMPT: Act as a launch strategist. Write three introductory social posts that: Explain who the partner is (their role, expertise, or value) Highlight why this collaboration serves my audience, especially busy, real-life marketers Gently build excitement without pressure or hype the tone must be friendly, helpful, and trust-building, fully aligned with the theme "Campaigns, Not Chaos". For each of the three posts, include: A clear focus (Post 1: Introduction, Post 2: Why This Partner Matters, Post 3: What This Means for the Audience)A short, encouraging message (30–50 words each) A subtle lead-in to future updates or launch content Partner Details: [Insert partner information] Audience: Busy marketers seeking clarity and simplicity Brand Voice: [Insert brand voice]

Step 3 – Write a 'What Questions Do You Have?' Email or Story: This builds connection and opens conversations.

PROMPT: Act as a customer-nurture specialist. Write either a short email or an Instagram Story script inviting people to ask questions about an upcoming collaboration. The tone must be calm, supportive, and approachable, aligned with "Campaigns, Not Chaos. "Please include: A gentle opening that acknowledges busy schedules and reassures the audience there's no pressure A simple explanation of the collaboration and why it might be helpful An open invitation for questions, framed in a way that feels safe and easy to respond to A short closing line that reinforces clarity, support, and accessibility Collaboration Details: [Insert partner/collaboration info] Audience: Busy marketers who value clarity and encouragement Brand Voice: [Insert brand voice] Format Preference: [Email or IG Story]

Reuse, Refine, repeat: Here's your real power: These procedures aren't one-and-done. They are reusable systems that grow with you. Save your best prompts. Tweak the ones that improve over time. Keep versions that got high engagement, strong email opens, or great responses. Your system becomes smarter, clearer, and more "you" every time you use it. You don't need: a new app, a complicated funnel builder, or a huge learning curve, you just need **one simple structure** you trust – and a willingness to return to it.

Executive Bonus Prompt: Team Collaboration Version

Team Collaboration PROMPT: Act as a team collaboration strategist supporting a small marketing team working toward consistent, organized campaigns. Using the notes or campaign idea below, create a clear, shared plan that aligns the whole team. Please include Role-Specific Guidance Write a short, actionable instruction for each role: Strategist: direction + core message Designer: visual requirements and brand cues, Copywriter: tone, hooks, and key points, Editor/Approver: review criteria and final checks, Social/Email Manager: scheduling and distribution notes, Cross-Functional Alignment Notes Highlight what each role needs from others (handoffs, dependencies, and shared files) Simple Multi-Approval Workflow Outline a 3–5 step approval sequence that keeps things moving without slowing the team down. Mini Content Calendar Provide a 5–7-day schedule with publishing order, platform notes, and must-hit messages. Brand Governance Reminders Include 3–5 quick rules to ensure the campaign stays consistent, calm, and on-message with "Campaigns, Not Chaos." Keep the overall plan clear, structured, and calming — designed to support busy teams who want coordination without complexity. Campaign Notes / Idea: [Paste your campaign notes] Team Roles: [Insert roles or names] Brand Voice: [Insert brand voice] Primary Goal: [Insert campaign goal] This is sustainable marketing. This is clarity over chaos. This is how real marketers create momentum in real life.

Closing Reflection

This chapter reinforced a defining choice: ease over overwhelm, clarity over chaos. Sustainable marketing is built around your reality, not around constant hustle. This is how momentum becomes sustainable, for the version of yourself who creates with confidence instead of strain.

Forward into chapter Six

Chapter Six focuses on how to integrate writing into real life in a way that feels natural and supportive. You will explore how to stay connected to your work as energy fluctuates, use everyday moments as creative fuel, and strengthen your voice with AI while keeping it authentically yours, even during demanding weeks, a reminder that your writing deserves to grow without pressure.

Chapter Six

Finding Your Writing Rhythm: A Gentle Pace That Supports You

Section One

Most marketers are taught that writing should follow a fixed schedule, a fixed output, and a fixed level of intensity. Write every day. Post three times a week. Stay consistent at all costs. When output drops, the assumption is simple: something must be wrong. This narrative is not only unrealistic — it is counterproductive. It ignores how cognitive energy, attention, and decision-making actually function in real life. Writing does not exist in isolation. It operates inside a human nervous system, within a real calendar, and alongside fluctuating energy demands. When those realities are ignored, writing becomes a source of pressure rather than a source of clarity or connection.

Writing is not a task to be enforced. It is a rhythm to be understood.

Your writing rhythm is the natural pattern through which ideas surface, energy rises and falls, and expression becomes possible without unnecessary strain. It is not something you impose through discipline alone. It is something you observe, respect, and design around. This awareness matters because writing energy is not the same as motivation. Motivation is emotional and inconsistent. Writing energy is physiological and contextual. It is shaped by sleep quality, cognitive load, emotional demand, environmental noise, recovery time, and mental clarity. On some days, the mind feels open and spacious. On others, it does not — and no amount of discipline changes that reality.

For many marketers, the issue has never been a lack of commitment. It has been the expectation that they should work against their own internal timing. This chapter challenges that expectation. It invites you to step away from rigid output standards and toward a writing approach that supports your capacity, your professional responsibilities, and your long-term creative sustainability — without compromising the quality or credibility of your marketing work. When marketing and content creation are aligned with rhythm, several shifts occur naturally. Writing becomes easier to initiate. Consistency becomes more stable. Voice becomes clearer and more confident over time. This is because consistency is no longer treated as a volume metric. It is reframed as a **structural outcome**. Your rhythm is not about producing content at a fixed rate. It is about designing a sustainable pattern for thinking, drafting, refining, and publishing that aligns with how your mind actually operates under changing conditions. When rhythm is present, output stabilises. When it is absent, even strong strategies and tools begin to break down under friction.

Most marketers work inside tight constraints: limited uninterrupted time, competing priorities, frequent context switching, and emotional labour tied to messaging and positioning. Under these conditions, output consistency — publishing at the same volume every week — becomes an unstable and often damaging measure. It creates pressure without improving effectiveness. Rhythm prioritises structure over volume. It allows output to fluctuate while preserving momentum and coherence. Writing rhythm is not about doing less. It is about removing unnecessary friction so the work you are already capable of can move forward with greater ease and reliability.

Section Two

Why Writing Resistance Appears (and Why It Is Misdiagnosed)

Many marketers delay writing because they assume it requires uninterrupted time, emotional readiness, or a completely clear head. In practice, waiting for those conditions often means waiting indefinitely. Effective writing systems are not built on perfect circumstances — they are built on **reliable entry points**, low friction, and clear next actions. Writing resistance is frequently mislabelled as a motivation problem. More often, it is a **cognitive overload signal**. Marketing writing is uniquely demanding because it requires multiple forms of thinking to happen at once: strategy, language selection, audience psychology, brand positioning, and performance considerations. When these demands stack, the brain defaults to avoidance — not because the marketer lacks discipline, but because the task begins with too many open decisions. When resistance appears, it is rarely "I don't feel like writing." It is more accurately: "I don't know where to start and starting feels energetically expensive."

Common triggers include Undefined objectives. Too much choice at the beginning of the task. Frequent context switching or fragmented attention. The blank page is not intimidating because it is empty. It is intimidating because it offers unlimited direction without constraint. When structure is missing, that weight compounds quickly. The solution is not more pressure. The solution is **structure**: small starts, reduced choice, and a clear definition of what "done" looks like. A well-designed prompt can remove friction at the exact moment resistance appears. Its purpose is not to generate ideas, but to **stabilise direction** and reduce the cognitive cost of starting.

Supportive Writing Prompt

PROMPT: Act as my writing partner. Here are my raw thoughts: [paste]. Help me turn them into one clear, grounded paragraph while preserving my voice. Do not add new ideas. Improve clarity, flow, and readability only. Tone: calm, professional, human.

Used correctly, this prompt keeps you writing with your thinking rather than fighting against it.

Designing Your Personal Writing Rhythm

A functional writing rhythm is **designed**, not discovered. It begins with a simple but often overlooked recognition: not all marketing work draws on the same mental resources. Most resistance emerges when high-cognition tasks are attempted without sufficient capacity. The solution is not to push harder, but to match the work to the conditions in which it can be completed effectively.

Step One: Classify Your Writing Tasks

Begin by grouping your marketing work into three cognitive categories:

High-cognition tasks: strategy, positioning, long-form drafting, narrative development

Medium-cognition tasks: editing, repurposing, refinement, optimisation, sequencing

Low-cognition tasks: formatting, organising, scheduling, light revisions

This classification introduces clarity before execution. It reduces friction by making the cognitive cost of each task visible.

Step Two: Map Tasks to Energy Patterns

Instead of assigning tasks to calendar slots, assign them to **energy windows**. **Ask:** When does my thinking feel most precise? When does language flow more easily? When is my attention fragmented or limited? This approach allows writing to occur when it is most efficient, rather than forcing demanding work into unsuitable conditions.

Step Three: Define Minimum Viable Output

One of the most effective rhythm stabilisers is defining what counts as completion. Momentum often collapses when every task is held to an unnecessarily high standard. Minimum viable output creates clear stopping points, reduces perfectionism, and accelerates completion cycles.

A draft completed at 70% quality is more valuable than a perfect idea that never leaves the planning stage. Rhythm reduces resistance by **pre-deciding** what would otherwise drain cognitive energy: What type of content comes next. What level of quality is required. How much time is reasonable to invest. When a draft is considered complete. When these decisions are made in advance, writing shifts from negotiation to execution. This is where the difference between **routine** and **rhythm** becomes critical.

Routines are rigid. They assume stable conditions: predictable time blocks, consistent energy levels, and uninterrupted focus. Marketing work rarely offers those conditions.

Rhythms are adaptive. They acknowledge variability and allow different types of work to coexist without forcing uniform output. Some days support strategic thinking. Some support refinement. Others support only maintenance — and all three have value. This is not about productivity tactics. It is about **task–energy alignment**. When high-cognitive work is forced into low-capacity moments, resistance increases. When tasks are matched to available energy, momentum becomes easier to sustain.

Designing a Writing Practice

Marketers lose rhythm not because they lack discipline, but because their systems are brittle. They lock deliverables to dates rather than readiness, treat all content as equal in cognitive demand, and penalise deviation instead of adapting to context. The result is a predictable cycle: missed deadlines create stress, stress increases resistance, and resistance leads to avoidance or rushed output. Over time, the system meant to support consistency becomes the source of its erosion. The most reliable way to stabilise a writing rhythm is to **lower the threshold for starting**.

Instead of asking yourself to "write a post," you capture a sentence, note an observation, save a question, record a voice memo, or draft a headline. These actions require minimal cognitive effort, yet they preserve momentum. These fragments are not incomplete work. They are **raw material**. Their value lies in continuity, not polish. Your responsibility is not to complete every idea in one sitting, but to keep the creative signal active. This is why capturing ideas consistently matters more than finishing drafts immediately. A rhythm-based approach reframes planning itself. It shifts focus from fixed programmes to content categories, from rigid scheduling to sequencing, and from punctuality to continuity. The objective is to **maintain motion**. When motion is preserved, completion follows naturally.

Identity-Based Consistency

Consistency becomes sustainable when it is anchored in **identity**, not performance. Marketers who define themselves as "inconsistent" tend to interpret disruption as failure. Those who see themselves as **adaptive** understand disruption as a normal feature of professional work.

This variance matters because performance fluctuates, but identity endures. Rhythm supports identity formation by creating repeatable, low-resistance behaviours: starting without excessive friction, completing work without exhaustion, and returning to writing without hesitation. Over time, these patterns reinforce confidence and reduce avoidance, not through pressure, but through reliability.

When marketers trust their rhythm, consistency stops being something to maintain and becomes something that naturally expresses how they work Consistency becomes sustainable when it is tied to identity rather than performance. Marketers who see themselves as "inconsistent" tend to interpret disruption as failure. Marketers who see themselves as "adaptive" interpret disruption as part of the process.

Section Three

AI as a Rhythm Stabiliser (Not a Replacement)

A functional writing practice is built on small entry points, low resistance, and clear next actions. This chapter introduces a professional writing model designed for busy marketers—one that integrates AI as a **support mechanism**, not a substitute for thinking. The most effective use of AI in writing is: You supply the intent, insight, or emotional truth. AI supply's structure, clarity, and options. One of the most common concerns marketers raise is the fear of losing their voice when using AI. The concern is understandable, but it usually stems from misuse. AI should not replace judgment or intent. It should **support execution**.

Used strategically, AI reduces activation energy—the effort required to begin. It performs best where momentum is fragile and capacity fluctuates. In practice, it excels at structuring starting points, clarifying direction, reducing cognitive load during drafting, and supporting refinement without emotional fatigue. This makes it particularly useful on low-energy days, during transitions between tasks, or when clarity exists but execution feels heavy. In this context, AI's role is **regulatory, not creative**. It smooths variability and stabilises output without replacing expertise. The most effective division of labour is straightforward: you supply the intent, insight, or emotional truth; AI supplies structure, clarity, and options. This balance preserves authenticity while reducing strain. For example, rather than forcing a polished paragraph when cognitive capacity is limited, you can offer something honest but unrefined—and ask AI to shape it, not invent it.

Supportive Writing Prompt

PROMPT: Act as my writing partner. Here are my raw thoughts: [paste]. Help me turn them into one clear, grounded paragraph while preserving my voice. Do not add new ideas. Improve clarity, flow, and readability only. Tone: calm, professional, human. Used this way, AI keeps you writing *with* your thinking rather than against it.

Sustainable Visibility Without Burnout

Burnout rarely results from volume alone. It emerges from **misalignment**—when output expectations are disconnected from capacity. In those conditions, messaging becomes mechanical and effortful, even when strategy is sound. When rhythm is respected, coherence is preserved across fluctuating output levels. Visibility does not depend on constant intensity; it depends on **recognisability**. Sustainable visibility is built on three fundamentals: continuity of presence, clarity of positioning, and consistency of tone. Together, these allow a brand to remain familiar and trustworthy without requiring relentless production. When marketing is aligned with rhythm, presence becomes stable rather than exhausting—and momentum is maintained without burnout.

The Rhythm Reset Framework

To operationalise rhythm, use a simple four-stage loop, designed to adapt to real-world conditions rather than resist them:

Reset — Identify current capacity and constraints

Align — Match tasks to available energy

Execute — Prioritise completion over optimisation

Review — Assess what flowed easily and what resisted

This framework is **cyclical, not linear**. It is intended to evolve with circumstances, allowing rhythm to be recalibrated without disruption or self-judgment. Writing does not need to be stationary. Some of the most effective marketing ideas emerge while walking, driving, completing routine tasks, resting, or waiting. The value lies not in where writing happens, but in **how ideas are captured and later developed**.

The writing-in-motion model consists of three stages:

Capture the idea immediately

Store it without judgment

Develop it later with AI support

This approach removes pressure from the moment of inspiration and shifts refinement to a time when energy and attention are better suited to the task. A single line captured at the right moment can later become a social caption, an email opening, a blog section, or a broader content theme. AI enables this expansion without requiring you to recreate the emotional or cognitive state in which the idea first appeared.

Low-Energy Writing Prompt

On days when capacity is limited, use the following prompt to preserve momentum without depletion:

PROMPT: I have limited energy today. Act as my writing partner and help me turn the idea below into a short, usable piece of writing. Keep it simple, clear, and aligned with my voice. Do not over-expand.
Idea: [insert]

Used appropriately, this approach keeps motion intact even when output must remain light.

Recognizing Progress Without External Validation

One of the most damaging assumptions in marketing is that progress only counts when it is visible or externally measurable. In practice, writing progress often appears internally long before it shows up publicly. Early indicators of meaningful progress include reduced resistance when starting, clearer articulation of ideas, faster decision-making, less self-editing during drafts, and a greater willingness to return to writing. These shifts are subtle, but they are structurally significant. They signal that your writing rhythm is stabilising.

To reinforce this progress, replace performance review with internal review.

At the end of a writing session, ask:

What moved forward today?

What felt easier than before?

What did I learn about my process?

This approach builds confidence through awareness rather than metrics. It reduces dependence on external validation and anchors consistency in observable process improvement.

Reflection Prompt

PROMPT: Help me reflect on today's writing session. Identify one thing that moved forward and one clear next step for tomorrow.

Used consistently, this keeps progress visible even when output remains light.

Why Chapter Six Is a Structural Turning Point

This is the point where your marketing practice stops being mechanical and becomes relational. You are no longer focused solely on producing content; you are learning how to remain present with your voice over time. That shift matters because marketing consistency is emotional before it is technical. Voice clarity determines trust. Sustainable pace prevents abandonment. A marketer who trusts their rhythm is far more effective than one who forces output.

The guiding question shifts from "*Do I feel like writing?*" to "*What kind of writing fits the energy I have today?*" That reframing becomes the foundation for everything that follows.

Closing Reflection

Marketing does not happen under ideal conditions. It unfolds alongside responsibility, unpredictability, and competing demands. A writing rhythm does not remove those pressures; it works around them. When rhythm is established, consistency becomes a by-product rather than a burden. Output stabilises, confidence strengthens, and marketing begins to support your work instead of draining it. This is not a productivity tactic. It is **professional sustainability**. Your writing no longer depends on perfect conditions. It depends on awareness, structure, and support. That shift is practical, durable, and repeatable.

Your writing no longer depends on perfect conditions. It depends on awareness, structure, and support. Your creativity does not need to survive on adrenaline.

Forward Into Chapter Seven

Chapter Seven is about seeing clearly. You will reflect on what you have built, recognise the systems now supporting you, and identify the confidence that has been forming beneath the surface. This is not a finish line. It is a recalibration.

Chapter Seven

The AI-Ready Content Calendar

Section One

You need a system that bends. A system that forgives interruptions, that keeps you moving even when your focus is low or your schedule changes by the hour. That's what an AI-ready content calendar gives you. It's not a rigid set of dates. It's a rhythm — a gentle, flexible structure that gets content out of your head and into motion. Let's build a version of a content calendar you can actually stick to.

Plan Themes, Not Posts

Instead of planning every single post, you choose 3–4 **weekly content themes** that guide your messaging. Themes give you clarity without pressure. Themes help you stay consistent without thinking too hard. Themes let AI do the heavy lifting. Here's a simple example:

Monday — Teach: Share a tip, insight, or short lesson.

Wednesday — Connect: Share a story, reflection, or behind-the-scenes moment.

Friday — Invite: Share a soft promotion, offer reminder, or invitation.

That's it. Three themes. Three pillars. Three touchpoints per week. Some weeks you'll post all three. Some weeks you'll post one. What matters is that you have a structure to return to, not a schedule to keep up with.

Why This Works for Busy Marketers

1. You always know what to create next. No more wondering, "What should I post?" Your themes tell you.

2. AI can easily generate content inside these themes. Themes = built-in

direction.

3. You avoid decision fatigue. Your brain relaxes when it knows the shape of the week.

4. It keeps you visible without burning out. Consistency becomes achievable.

5. It works even when your life is unpredictable. Themes flex with you.

Your AI-Ready Weekly Calendar Template

Monday – Teach Something Valuable: This positions you as knowledgeable, helpful, and trustworthy.

PROMPT: Act as a content educator. Turn the idea below into a short teaching post that fits easily into an AI-ready content calendar. Please include: A clear, simple explanation of the topic in everyday language One practical takeaway that readers can use immediately A warm, approachable tone that makes the lesson feel easy, not overwhelming A concise structure suitable for social media, newsletters, or content libraries Keep the post brief, useful, and reader friendly. Topic: [Insert topic] Audience: [Insert audience] Brand Voice (optional): [Insert brand voice]

Wednesday – Share a story: Stories build connection. Behind-the-scenes builds trust.

PROMPT: Act as my storytelling partner. Turn the experience below into a relatable, warm post that fits seamlessly into an AI-ready content calendar. Please include: A human, approachable opening that helps readers see themselves in the moment A simple narrative flow that connects the experience to a broader theme or insight A warm, grounded tone that feels real, supportive, and easy to engage with A soft closing line that invites reflection or conversation without pressure Keep the post concise, reader-friendly, and suitable for social platforms or newsletters. Experience or Thought: [Describe event or moment] Audience: [Insert audience] Brand Voice (optional): [Insert brand voice

Friday – Extend an Invitation (Soft Promotion): No pressure. No hard sales. Just clarity and confidence.

PROMPT: Act as a gentle marketer. Write a soft promotional post about [offer/book/product]. The tone must be warm, friendly, and honest, suitable for an AI-ready content calendar that values trust and consistency. Please include: A welcoming opening line that feels human and low-pressure One clear benefit or reason the offer might help someone A gentle, honest explanation of what the offer is or why it exists A soft call-to-action that invites curiosity without pushing for a sale Keep the post short, grounded, and easy to read, suitable for social media or newsletters. Offer/Book/Product: [Insert details] Audience: [Insert audience] Brand Voice (optional): [Insert brand voice]

Optional: Weekend – Reflect or Celebrate Wins: If you have the energy, share highlights, gratitude, or reader feedback.

PROMPT: Act as a positive reinforcement coach. Turn the weekly wins listed below into a short reflection post that supports an AI-ready content calendar. Please include: A warm opening line that celebrates progress without pressure A brief, uplifting reflection that connects the wins to growth, learning, or momentum A supportive, encouraging tone that reinforces consistency and confidence A simple closing line that invites readers to notice their own wins Keep the post short, genuine, and easy to read. Weekly Wins: [Insert wins here] Audience: [Insert audience] Brand Voice (optional): [Insert brand voice]

How to Use AI to Fill Your Calendar in Under 20 Minutes

This is where your system becomes powerful.

Step 1 – Choose your themes: (Teach, Connect, Invite – or your preferred three)

Step 2 – Create 3 AI prompts per theme: Use the prompts above.

Step 3 – Ask AI for multiple versions: This gives you 9–12 posts in minutes.

Step 4 – Drop them into your Google Doc or planner: No scheduling pressure – just organized options.

Step 5 – Use when ready: Post on days that suit your real life.

This removes the weekday stress and gives you content you can pull from anytime you need it.

Monday – Teach:

PROMPT: Act as a content educator. Turn the book excerpt below into a clear, concise teaching post that helps readers understand one core idea. Please include: A simple, direct explanation of the main concept One practical takeaway readers can apply immediately A warm, helpful tone that supports learning without overwhelm A clean structure suitable for social media, newsletters, or educational posts Keep the post short, useful, and easy to read. Book Excerpt: [Paste excerpt here] Audience: [Insert audience] Brand Voice (optional): [Insert brand voice]

Wednesday – Connect:

PROMPT: Act as a storytelling partner. Turn the story about why I wrote this book into a warm, personal post that helps readers feel connected to me and the deeper purpose behind the work. Please include: A relatable opening that eases readers into the story A personal moment or insight that explains why the book exists A warm, human tone that feels genuine, grounded, and honest A simple closing line that invites reflection or connection without pressure Keep the post short, heartfelt, and easy to read, suitable for a mid-week connection post inside an AI-ready content calendar. Story: [Paste your story here] Audience: [Insert audience] Brand Voice (optional): [Insert brand voice]

Friday – Invite

PROMPT: Act as a gentle marketer. Write a soft promotional post explaining who this book is for and why it will help them. Please include: A warm, welcoming opening that speaks directly to the reader's real-life challenges A simple explanation of who will benefit most from the book One clear value or outcome the book helps them achieve A soft, respectful invitation to learn more — without pressure or hype Keep the tone friendly, supportive, and ease-focused, aligned with an AI-ready content calendar's end-of-week rhythm. Book Title: [Insert book title] Audience: [Insert audience] Brand Voice (optional): [Insert brand voice]

This gives you a ready-to-go flow for your book, your brand, and your marketing presence. This is the calm marketer's calendar — designed for real people with real responsibilities.

Section Two

Turning Your Calendar into a Monthly AI Workflow

A weekly rhythm gives you stability. But a monthly workflow gives you momentum — the kind that quietly builds your visibility, supports your launches, and frees your mental space. A monthly AI workflow protects your energy, reduces decision fatigue, and helps you show up consistently even when life pulls you sideways.

The 4-Part Monthly Workflow (Simple, Flexible & AI-Supported)

You don't need a marketing department. You don't need long planning sessions. You need **four touchpoints** each month — each one supported by AI.

1. Month Kick-Off: Set Your Focus (10–15 minutes) Every month starts with one question: "What do I want my audience to understand or feel this month?" That's your anchor. It might be clarity, confidence. consistency, a specific problem you help solve, a book launch, a seasonal offer, a mindset shift, a simple teaching theme, you don't need ten goals. You need *one* focus.

PROMPT: Act as my marketing strategist. Based on this theme: [insert theme], create a simple monthly focus statement in one or two sentences. Please ensure the statement: Clarifies the core priority for the month. Sets a calm, achievable direction that's easy for a busy marketer to follow. Feels supportive, grounded, and intentional. Aligns naturally with an AI-ready content calendar rhythm. Keep the tone confident, warm, and steady, and ensure the statement is short enough to guide decisions without adding pressure. Theme: [Insert theme] Audience: [Insert audience] Brand Voice (optional): [Insert brand voice]

2. Build Your Monthly Post Bank (15 minutes): Using your weekly themes (Teach, Connect, Invite), create a pool of 8–12 posts. These aren't tied to dates. They're simply ready when needed.

PROMPT: Act as a content creator. Using my monthly focus and the weekly themes — Teach, Connect, Invite — create four posts for each theme. Please ensure: Teach Posts (4): Clear, simple lessons based on the monthly focus One practical takeaway per post Tone: warm, helpful, easy to understand Connect Posts (4): Personal insights, stories, or reflections tied to the focus Relatable moments that build trust Tone: human, warm, and supportive Invite Posts (4): Soft, gentle invitations to explore a product, offer, or idea One clear benefit per post Tone: calm, encouraging, pressure-free Across all posts: Keep the tone warm, practical, and supportive Ensure they are short, clear, and content-calendar-ready Align everything with the monthly focus Monthly Focus: [Insert monthly focus] Audience: [Insert audience] Brand Voice (optional): [Insert brand voice]

This gives you 12 posts in minutes.

3. Plan Your Monthly "Cornerstone Content" (10–20 minutes)**:** Once a month, create one deeper piece of content: a blog, a newsletter, a video, a teaching carousel, a chapter excerpt, a case study, a behind-the-scenes reflection. This becomes the *anchor* your other posts can repurpose.

PROMPT: Act as a content strategist. Create one piece of cornerstone content based on this monthly theme: [insert theme]. Format: [blog / email / video script / carousel] Tone: conversational and supportive Please ensure the cornerstone content includes: A clear, engaging introduction that frames the theme in everyday, relatable terms Three to five key points that teach, guide, or support the audience Simple, conversational explanations that reduce overwhelm and increase clarity A supportive closing message that encourages the reader/viewer without pressure A natural flow that feels accessible and helpful for busy marketers Keep the content structured, easy to follow, and aligned with the monthly theme so it can serve as the anchor piece for the rest of the AI-ready content calendar. Monthly Theme: [Insert theme] Audience: [Insert audience] Brand Voice (optional): [Insert brand voice]

4. Run a Monthly Engagement Loop (5–10 minutes per week). Engagement isn't about spending hours online. It's about showing up with *intention.* Your monthly loop includes replying to comments, sending thank-you messages, responding to DMs, asking a question, sharing a reflection, reposting a review, celebrating a small win. AI can help you create responses that feel warm and human without requiring emotional energy you may not have.

PROMPT: Act as a warm, friendly engagement assistant. Write five variations of supportive, human comments I can use to respond to my audience across social platforms. Please ensure each comment: Sounds genuine and human, not scripted or generic Uses a tone that is warm, encouraging, and approachable Can be applied flexibly to a variety of audience posts Helps strengthen connection and conversation Fits naturally within an AI-ready content calendar focused on consistency and community Keep all five responses short, uplifting, and easy to personalize. Audience Context: [Insert details about your audience] Brand Voice (optional): [Insert brand voice]

Putting It All Together: Your Monthly Workflow in Motion Here's what it looks like in real life

Week 1 Create monthly focus, generate your post bank, Publish cornerstone content

Week 2 Post 1–2 items from your bank. Light engagement

Week 3 Post 1–2 more items. Share something personal or behind-the-scenes. Light engagement

Week 4 Share one soft invitation (offer or book mention) Review what worked. Save successful prompts Reset for next month. Even if you are only posting twice a week, you stay visible, reliable, helpful, calm and fully in charge of your message This is the opposite of hustle culture. This is sustainable content creation for real people with full lives.

Section Three

Seasonal Campaign Planning with AI (Holidays, Launches, Sales)

Seasonal campaigns can feel overwhelming – not because they are complicated, but because they often arrive during the busiest, most emotionally loaded times of the year. Holidays. School breaks. Quarter deadlines. Family commitments. Unexpected life interruptions. A seasonal campaign doesn't need to be big. It needs to be clear, timely, and relevant – and AI can help you prepare it in a fraction of the time. Whether it's Christmas, Mother's Day, EOFY sales, back-to-school season, or the launch of your book, you can use the same structure to plan something powerful, simple, and achievable.

Step 1 – Identify the Seasonal Angle (Simple & Clear) You don't need 20 ideas. You need *one clear angle* that ties your message to the seasonal moment.

Ask yourself: What does my audience need right now? How can my book/product/service support them during this season? What problem or emotion is strongest at this time of year? What theme naturally matches my message?

Examples

Christmas - simplicity, support, inspiration, gift ideas.

Back-to-school -organization, clarity, routines

EOFY ▯ productivity, planning, systems

New Year ▯ fresh start, new habits, vision

PROMPT: Act as a seasonal marketing strategist. Given this season: [insert season or moment], suggest five simple angles I can use to connect my message or book to what my audience cares about right now. Please ensure the angles: Reflect real seasonal emotions or priorities (energy, challenges, mood, pace, celebrations, transitions) Tie naturally back to my message or book, without feeling forced Are simple, practical, and immediately usable in an AI-ready content calendar Speak directly to busy, real-life marketers who want clarity and relevance Offer variety, such as storytelling angles, teaching angles, promotional angles, or reflective angles Keep the tone warm, timely, and supportive, with a focus on meeting the audience where they are. Season: [Insert season or moment] Message / Book: [Insert title or focus] Audience: [Insert audience] Brand Voice (optional): [Insert brand voice]

Step 2 – Create a 3-Part Seasonal Content Plan. Seasonal campaigns don't need complexity. They need structure. A 3-part framework works beautifully:

1. Awareness Post:

PROMPT: Act as a seasonal content strategist. Write an awareness post that explains the seasonal challenge or emotion my audience is likely experiencing right now. Please ensure the post: Names the seasonal challenge or emotion clearly (e.g., burnout, renewal, pressure, slow pace, fresh start, overwhelm) Acknowledges the reality of how this season affects energy, focus, or motivation Uses a warm, steady, empathetic tone that helps readers feel seen and understood Stays short, relatable, and calm, suitable for an AI-ready content calendar Connects gently to my broader message or book without promoting Keep the focus on awareness, empathy, and connection, not solutions or selling. Seasonal Challenge/Emotion: [Insert here] Audience: [Insert audience] Brand Voice (optional): [Insert brand voice]

2. Value Post:

PROMPT: Act as a seasonal content educator. Write a value post that offers tips, support, or insight related to the current season. Please ensure the post: Provides 1–3 simple, practical tips that match the seasonal mood, pace, or challenges Uses a warm, supportive tone that makes readers feel guided, not pressured Includes one insight or perspective that helps the audience feel more grounded or capable Stays short, clear, and easy to apply, fitting seamlessly into an AI-ready content calendar Connects gently to my larger message or book without shifting into promotion Keep the focus on helping, encouraging, and supporting the reader based on what the season brings. Season: [Insert season] Audience: [Insert audience] Brand Voice (optional): [Insert brand voice]

3. Invitation Post:

PROMPT: Act as a gentle, human-focused marketer. Write an invitation post that shares my offer/book/service in a warm, human way. Please ensure the post: Opens with empathy or understanding, acknowledging what the audience may be feeling this season Explains the offer/book/service in clear, simple language Highlights one meaningful benefit that truly supports the reader Uses a soft, respectful tone — no pressure, no hype Ends with a gentle invitation to explore or learn more, allowing the reader to choose Keep the post short, calm, and supportive, aligning with the rhythm of an AI-ready content calendar. Offer/Book/Service: [Insert details] Audience: [Insert audience] Season (optional): [Insert season] Brand Voice (optional): [Insert brand voice]

This feels natural and supportive — never pushy.

PROMPT: Act as a content strategist. Using this seasonal angle: [insert], create a 3-part content plan that includes: An Awareness Post Name the seasonal feeling, challenge, or shift Use empathy and recognition to help the audience feel seen A Value Post Offer simple, practical tips or insight connected to the season Keep the tone supportive and easy to apply A Soft Invitation Share my offer/book/service in a warm, human way Highlight one meaningful benefit and end with a gentle call to explore more Tone: supportive, human, and practical, aligned with an AI-ready content calendar. Please keep all three posts short, clear, and grounded, suitable for busy readers. Seasonal Angle: [Insert angle] Offer/Book/Service: [Insert details] Audience: [Insert audience] Brand Voice (optional): [Insert brand voice]

Step 3 – Turn the Plan into Actual Posts. Once the structure is set, AI handles the heavy lifting.

PROMPT: Act as a content creator. Turn the 3-part seasonal content plan below into full captions ready to publish. Tone: warm, conversational, and supportive – suitable for busy readers and aligned with an AI-ready content calendar. For each caption: Expand the idea with clear, human-centered language Add one relatable or empathetic line to deepen connection Ensure the message stays simple, grounded, and easy to read Keep each caption brief but complete, suitable for Instagram, Facebook, or LinkedIn Preserve the sequence: Awareness □ Value □ Soft Invitation 3-Part Seasonal Plan: Awareness: [Insert summary] Value: [Insert summary] Soft Invitation: [Insert summary] Offer/Book/Service: [Insert details] Audience: [Insert audience] Brand Voice (optional): [Insert brand voice]

Step 4 – Create a Simple Seasonal Email: Seasonal emails don't need to be long. In fact, short, warm, and human often performs best during busy months. Include: one seasonal insight, one helpful reminder, one invitation, one link

PROMPT: Act as an email copywriter. Write a short, friendly seasonal email based on this angle: [insert seasonal angle]. Please include: One relatable seasonal insight that helps the reader feel understood One simple, practical tip they can use right now One gentle call to action aligned with a supportive, no-pressure tone A voice that is warm, conversational, and easy to read, suitable for an AI-ready content calendar Keep the email brief, human, and helpful, acknowledging that my audience is busy and appreciates clarity over complexity. Seasonal Angle: [Insert]

Audience: [Insert audience] Brand Voice (optional): [Insert brand voice] Offer/Book/Service (optional): [Insert details]

Step 5 – Add a DM or Private Message Script (Optional but Powerful) People often respond more positively to personal communication during seasonal periods.

PROMPT: Act as a relationship-based marketer. Write a warm, simple DM I can send to someone who engages with my seasonal posts. Please ensure the message: Acknowledges their engagement in a genuine, human way Uses a warm, conversational tone that feels like a real person, not a script Offers one small point of connection related to the seasonal angle Has no pressure, just an easy invitation to continue the conversation or explore further Is short enough to feel natural in a DM, while still being supportive and friendly Keep the tone aligned with an AI-ready content calendar: calm, human, and community-focused. Seasonal Angle: [Insert seasonal theme] Audience: [Insert audience] Brand Voice (optional): [Insert brand voice] Offer/Book/Service (optional): [Insert details]

Step 6 – Build a Seasonal Offer (If Relevant) This step applies when you want to: offer a book bundle run a limited-time discount, sell a workshop, promote a digital product, run a giveaway, Keep it simple.

PROMPT: Act as a marketing strategist. Help me design a simple seasonal offer for [insert product/book/service]. Please include Offer Description A short, clear explanation of what the seasonal offer includes. Who It's for A concise statement describing the audience that will benefit most during this season. Key Benefit One meaningful, seasonally relevant benefit that helps busy people understand why this matters now. One-Sentence Call-to-Action Warm, simple, and low-pressure – aligned with the tone of an AI-ready content calendar. Tone: supportive, practical, and human, with a focus on helping the audience feel guided rather than sold to. Product/Book/Service: [Insert details] Season: [Insert season or theme] Audience: [Insert audience] Brand Voice (optional): [Insert brand voice]

Step 7 – Repurpose Everything into Multiple Formats. Seasonal content is perfect for repurposing. Turn one idea into: a caption, a reel script, a carousel, a newsletter intro, a story, a short video, a DM reply, a book promotion snippet

PROMPT: Act as a repurposing expert. Using the seasonal content below, turn it into the following assets: A reel script - Short, supportive, and easy to voice - Includes hook, main point, and soft close A carousel outline - 5–7 slides - Clear structure (hook □ insight □ value □ supportive message □ takeaway) A tweet (or micro-post) - One sentence - Punchy, empathetic, seasonally aligned An email subject line - Clear, curiosity-driven, supportive - 5–7 words A short promo message - Warm, low-pressure - One benefit + gentle CTA. Tone for all assets: supportive, aligned, and human, consistent with an AI-ready content calendar. Seasonal Content: [Paste here] Audience: [Insert audience] Brand Voice (optional): [Insert brand voice] Offer/Book/Service (optional): [Insert details]

You now have a seasonal strategy that: adapts to your time, respects your energy, leverages your message keeps your marketing grounded and humane, never feels rushed or complicated. And — most importantly — it gives you a reliable structure you can use every holiday, every campaign, every launch. AI becomes your seasonal support partner, helping you stay visible without stress.

Closing Reflection

This chapter reframed planning as support rather than pressure. Instead of rigid calendars and unrealistic schedules, you designed a system that adapts to real life. By working with themes, trusted structures, and AI support, you created monthly content from clarity rather than urgency. Your calendar is no longer a source of guilt. It is a guide that keeps you moving forward, even when life becomes busy. The foundation is already in place. What remains is simply following your own rhythm.

Forward into Chapter Eight

With a content calendar you can trust, the next step is building systems that carry the workload when capacity is limited. Chapter Eight focuses on automation, templates, and workflows that save time without sacrificing quality. The goal is not more effort, but lighter execution. By the end of the chapter, you will have structures that allow you to create confidently, consistently, and calmly—even during your busiest weeks

Chapter Eight

Automation, Templates & Systems That Save You Hours

Section One

The AI Systems that: take pressure off your mind, support your creativity, reduce decision fatigue, help you stay consistent, and free hours of your week build once and use forever — templates, workflows, and automations that save us time again and again.

No longer creating content from scratch or rewriting the same email in ten different ways. with Ai's help we can have systems that: generate, sort, refine, personalize, and deliver...without draining our energy.

Systems that AI helps you build in minutes.

1. The Reusable Engineered Prompt Library (Your Always-Ready Assistant). You've built the foundation, but here's where it becomes a working system. A good prompt library isn't static — it evolves with you. You add prompts that performed well, prompts that saved time, prompts that improved your writing, prompts that broke writer's block, prompts that made launching easier. And you use them again and again — effortlessly.

Prompt to grow your library

PROMPT: Act as my professional prompt engineer. Review the prompts I've provided and redesign them for maximum clarity, precision, and consistency. Strengthen the structure, eliminate ambiguity, and ensure each prompt produces repeatable, high-quality outputs. Deliver improved versions that follow best-practice prompt engineering principles and maintain alignment with my brand voice and audience. Here are the prompts to refine: [paste prompts].

2. The Email Template Vault (Stop Rewriting Everything) Most marketers write the same types of emails repeatedly: welcome emails, launch announcements, soft reminders, thank-you messages, follow-ups; book promo emails, newsletter intros. Instead of rewriting these from scratch, build an email template vault. Just three templates can save you dozens of hours per year.

Prompt to create your vault

PROMPT: Act as an email strategist. Create three evergreen email templates I can reuse at any time: Welcome Email - Warm, friendly introduction - Brief explanation of what subscribers can expect - One simple next step Launch Announcement - Clear explanation of what's launching - Who it helps and the key benefit - Supportive, low-pressure tone - Gentle call to action Soft Reminder Email - Short and human - One reminder benefit - Respectful, easy-to-decline tone - Closing line that feels supportive Tone for all three: friendly, simple, and human — aligned with systems that save time and reduce overwhelm. Please ensure each template is: Reusable with light edits Clear, skim-friendly, and audience-focused Written with emotional warmth and professional simplicity Audience: [Insert audience] Brand Voice (optional): [Insert brand voice] Offer/Book/Service (optional): [Insert details]

3. Social Caption Templates (The Cure for Blank Screens) Imagine having 10–15 caption starters ready at any moment. That's what a caption template system gives you.

These templates remove mental friction on tough days.

Engineered Prompt to build them

PROMPT: Act as a social media copywriter. Create 15 reusable caption templates I can use for education, connection, and soft promotion. Please deliver: 1. Education Templates (5) Clear teaching structure One insight or tip per template Easy to adapt for different topics Tone: warm, practical, and supportive 2. Connection Templates (5) Personal storytelling or reflective prompts Helps deepen relationship with the audience Gentle, conversational tone Short and easy to customise 3. Soft Promotion Templates (5) Low-pressure promotional structure One clear benefit Warm, honest tone Gentle call to action Each template should be: Flexible (usable across industries) Skim-friendly and easy to personalize Suitable for automated or batched content workflows Consistent with a human, supportive brand voice Audience: [Insert audience] Brand Voice (optional): [Insert brand voice] Offer/Book/Service (optional): [Insert details]

4. Repurposing Automations (Your Time Multiplier) One piece of content should become six — without more effort, AI lets you turn: blog □ caption □ email □ reel script □ carousel □ DM reply, book excerpt □ newsletter □ teaser □ story, chapter idea □ Q&A post □ reflection □ hook list. This is how you expand your reach without expanding your workload.

PROMPT: Act as a repurposing expert. Using the content below, generate five aligned content assets: A caption- Short, warm, and easy to read- Includes one clear idea carousel outline- 5-7 slides- Structure: hook □ insight □ value □ takeaway □ soft close A newsletter paragraph- Supportive, helpful tone- Expands the idea with clarity and calm- One practical insight A short video script- 15-25 seconds- Conversational tone- Clear hook, core message, and soft sign-off A DM reply- Warm, simple, human- Acknowledge the person + offer one relevant next step Tone across all outputs must be consistent, warm, and helpful, suitable for automated content workflows and reusable templates. Source Content: [Paste content here] Audience: [Insert audience] Brand Voice (optional): [Insert brand voice] Offer/Book/Service (optional): [Insert details]

5. The 'Evergreen Offer' System (Always Ready to Promote) Every marketer — every author — needs 2-3 evergreen messages ready to go: why your offer/book matters, who it's perfect for, what transformation it provides.

These are not hype messages. They're support messages — guiding your audience with calm clarity.

PROMPT: Act as a messaging strategist. Create three evergreen messages I can use anytime to describe my offer/book. Please ensure each evergreen message: Clearly explains what the offer/book is States who it's for in a simple, relatable way Highlights one core benefit or outcome Is short, flexible, and reusable across social captions, emails, websites, or DMs Uses a tone that is warm, confident, and helpful, aligned with systems that save time and reduce overwhelm Produce three variations with slightly different angles, such as: Clarity-focused message Benefit-focused message Problem-solving message Each should be strong enough to stand alone and easy to personalize for different contexts. Offer/Book: [Insert details] Audience: [Insert audience] Brand Voice (optional): [Insert brand voice]

6. AI-Assisted Weekly Planning (10 Minutes, Zero Stress) This is where your templates, prompts, and automations work together. Each week, ask AI

PROMPT: Act as my content planner. Using my templates and themes, create a simple weekly plan that includes: Three posts - One teaching or educational post - One connection or storytelling post - One soft promotional post One email idea - Warm, practical, and aligned with the week's focus One repurposing suggestion - A quick, efficient way to reuse one of the weekly pieces across another platform or format Please ensure the plan: Aligns with my existing templates and content themes Is easy to follow, even on busy weeks Uses a tone that is warm, practical, and supportive Helps maintain a consistent presence with minimal effort Keep the structure clear, simple, and ready to implement within an automated system. Templates & Themes: [Insert details or paste your themes] Audience: [Insert audience] Brand Voice (optional): [Insert brand voice] Offer/Book/Service (optional): [Insert details]

Ten minutes. One rhythm. A whole week ready

Why These Systems Matter: Because we are: juggling responsibilities, caring for loved ones, managing work and colleagues, balancing our energy, navigating real-world demands, we don't need more work, we need systems that carry the load for us. A reusable template is not a shortcut —

it's a form of self-support. A prompt library is not a tool – it's a space to restart when you're tired. Repurposing isn't cheating – it's working smarter with what we have already created. AI is protecting our energy so we can use our voice where it matters. These systems help us stay consistent, confident, and grounded – no matter what season of life we're in.

Section Two

Automations That Quiet Your Mind and Simplify Your Week

If you've ever felt mentally crowded from trying to remember every task, every idea, every follow-up, every post, every message – you're not alone. Marketing isn't hard because the work is complicated. Marketing is hard because it demands our attention even on days when attention is in short supply. That's why light, thoughtful automation is a gift – not a gimmick. These automations handle the predictable parts of marketing so you can save your energy for the work that truly needs your voice, your story, your heart.

1. The Idea Capture Automation (Never Lose a Thought Again): Ideas come at the most inconvenient moments: while driving, while watching TV, in the garden, during dinner, at 10pm when your brain finally unwinds. You don't need a perfect system – you need a catch-all.

Use: a notes app, a voice memo, a Google form, a WhatsApp chat with yourself, a dedicated inbox folder or even a physical notebook, then ask AI to turn those rough fragments into structured content later.

PROMPT: Act as my clarity assistant. Turn the raw notes below into three organized content ideas I can use this week. For each content idea, include: A clear title A one-sentence summary of the angle or message A recommended format (post, email, reel, carousel, etc.) One practical takeaway or point of value for the audience Please ensure the final ideas are: Easy to implement Aligned with my templates and content themes Warm, supportive, and practical Suitable for use in an automated or templated content system Keep the structure simple, encouraging, and ready to use. Raw Notes: [Paste notes here] Audience: [Insert audience] Brand Voice (optional): [Insert brand voice] This simple automation relieves mental pressure and gives your ideas a safe home.

2. The Weekly Content Generation Automation (Your 10-Minute Reset)

Once a week, run a single automation that creates your content foundation. You only need two ingredients: your 3 weekly themes your monthly focus. Then AI creates your content for the week.

PROMPT: Act as my weekly content planner. Using: My monthly focus: [insert] My weekly themes: Teach, Connect, Invite Generate the following: Three ready-to-post captions Teach: A short educational post that offers one actionable insight Connect: A warm, personal or reflective post that builds trust Invite: A soft promotional post with one clear benefit and a gentle CTA One email idea. Warm, grounded, and aligned with the monthly focus Provide a brief description of the angle One repurposing suggestion Choose one of the week's captions Suggest how to turn it into another format (reel, carousel, tweet, etc.) Tone for all outputs: warm, grounded, and human, suitable for busy marketers using automated or templated systems. Please ensure the plan is: Easy to implement Consistent with the weekly rhythm and monthly theme Supportive and clear Immediately usable Monthly Focus: [Insert] Audience: [Insert audience] Brand Voice (optional): [Insert brand voice]

3. The Content Repurposing Automation (Work Once, Use Often): Every piece of content you create should serve you in five other places — without extra work. You speak once. AI does the multiplying.

PROMPT: Act as a content repurposing expert. Using the single piece of content below, create the following assets: An Instagram caption – Short, engaging, conversational – One core idea + soft close A LinkedIn post – Clear, confident, professional – Expands the idea with one insight or example – 120–180 words A short email – Warm, supportive tone – One takeaway + gentle call to action – Under 120 words A 5-slide carousel outline - Slide 1: Hook - Slide 2: Core insight - Slide 3: Supporting point - Slide 4: Practical takeaway - Slide 5: Soft invitation A 20-second video script - Hook (3 seconds) - Core message (12 seconds) - Supportive close or CTA (5 seconds) - Conversational, human, easy to speak aloud Tone across all formats must be consistent and aligned with my voice—warm, clear, supportive, and grounded. Source Content: [Paste your content here] Audience: [Insert audience] Brand Voice: [Insert brand voice]

This automation alone can save you dozens of hours each month

4. The Scheduling Automation (Set It and Forget It): If posting consistently feels overwhelming, use a scheduler: Meta Business Suite (free), Buffer, Later, SocialBee, Canva scheduler. You can prepare everything in one calm session and let the scheduler handle the rest.

Prepare scheduled content

PROMPT: Act as my scheduling assistant. Using the six pieces of content below, create a posting schedule that spreads them naturally across the next two weeks. Please ensure the schedule: Feels balanced and human, not overwhelming Follows a natural rhythm — alternating education, connection, and soft promotion Includes platform suggestions (IG, FB, LinkedIn, email, video, etc.) Recommends optimal posting days for visibility and audience energy Leaves space for real life — no rigid daily posting unless content fits Reflects a tone of calm, clarity, and sustainability, aligned with systems that save time Format the schedule as: Week 1 (Days + Content Type + Platform) Week 2 (Days + Content Type + Platform) Include one optional repurposing touchpoint per week Keep it simple, warm, and easy to implement. Six Content Pieces:[Insert][Insert] [Insert] [Insert] [Insert] [Insert] Audience: [Insert audience] Brand Voice (optional): [Insert brand voice] Platforms (optional): [Insert preferred platforms]

This keeps your marketing alive even on difficult weeks

5. The Follow-Up Automation (The Quiet Sales Engine): Most sales come from follow-up — not first contact. But following up manually drains your emotional and mental energy. AI can write warm, human, soft follow-ups that don't feel pushy.

PROMPT: Act as a relationship-based marketer. Write five friendly follow-up messages I can use for people who have: engaged with a post requested more info clicked a link Tone must be gentle, supportive, and zero pressure — warm enough to build trust, simple enough to use in automated or manual follow-ups, and fully aligned with human-first communication. Each message should: Acknowledge their action (like, comment, click, or request) Offer one small point of support or clarity Use warm, conversational language Include a no-pressure next step (optional to explore, ask a question, or reply) Stay short, human, and easy to personalize Context or Offer: [Insert details] Audience: [Insert audience] Brand Voice (optional): [Insert brand voice]

6. The Review & Reflection Automation (Your Progress Booster): Reflection is where clarity grows — but most people skip it because it feels like "extra work. "Let AI summaries your wins for you.

PROMPT: Act as my progress coach. Here are the things I posted and worked on this month: [paste list]. Please: Summaries my progress in a clear, encouraging way- Highlight consistency, creativity, and momentum- Note any patterns or improvements Identify what worked best- Content types, formats, themes, or actions that performed well- Insights based on tone, timing, or audience engagement Give me three focus areas for next month- Clear, practical, and achievable- Aligned with automation, templates, and systems that save time- Supportive, not overwhelming Tone: warm, grounded, and supportive, helping me stay motivated while refining my system. Keep the summary and recommendations short, actionable, and easy to implement in an AI-ready workflow. Monthly Activity List: [Paste list here] Audience: [Insert audience] Brand Voice (optional): [Insert brand voice]

Why These Automations Matter: Because they protect the most precious parts of your day: your energy, your focus, your emotional bandwidth, your time, your creativity You don't need to be "on" all the time. You don't need to force yourself to create when you're tired. You don't need to carry everything in your head. You need systems that think with you, categorize for you, and prepare for you — so you can show up with clarity and heart when it matters most. That's the real promise of AI: a quieter mind, a lighter workload, and a marketing rhythm that finally feels sustainable.

Section Three

Templates You Can Copy, Paste, and Use Forever

Templates are more than shortcuts. They are foundations — the starting point that steadies your hand, quiets your mind, and gives you something strong to build on. You need a sentence starter. A structure. A gentle nudge in the right direction. That's what these templates give you: clarity, speed, and a sense of calm.

Template 1: The "Teach Something Valuable" Caption. Use this when you want to share insight, educate, or position yourself as helpful—not preachy.

TEMPLATE: Opening (Hook): Here's something many people don't realise about [topic]...Insight (Teach Simply): The truth is, [insert clear, practical insight]. It matters because [explain why this helps your audience]. Value (Make It Actionable): If you want to make this easier, try [one simple tip, step, or shift]. It's a small change that creates a big difference over time. Close (Warm, Non-Preachy): Hope this helps you feel a little more confident with [topic]. What part of this stands out for you? Tone: Helpful, approachable, non-preachy Best for: Educational posts, value-driven content, positioning yourself as a supportive guide Platforms: Instagram, Facebook, LinkedIn, newsletters

PROMPT: Act as a social media educator. Write a short caption that teaches my audience one practical, valuable insight about [insert topic]. Please ensure the caption includes: A helpful, approachable opening line that feels human and non-preachy One clear piece of practical insight the audience can apply today A single, easy-to-remember take-away A light CTA that encourages reflection or conversation (not sales or pressure) Tone: helpful, approachable, and non-preachy — aligned with automation-friendly systems and content that feels warm and trustworthy. Keep the caption short, clear, and ready to post. Topic: [Insert topic] Audience: [Insert audience] Brand Voice (optional): [Insert brand voice]

Why It Works: Clear problem, Simple solution, encouraging tone, Builds authority gently

Template 2: The Storytelling Connection Post: Use this to build trust, share vulnerability, or reveal a behind-the-scenes moment.

TEMPLATE: Here's something I realized recently: [insert insight]. It came from [brief experience / moment / challenge], and it reminded me how often we're all navigating versions of the same thing. If you've ever felt the same, you're not alone — and here's what helped me: [insert simple lesson, shift, or takeaway]. Tone: warm, grounded, supportive Best for: connection posts, behind-the-scenes moments, gentle storytelling, weekly reflections Platforms: Instagram, Facebook, LinkedIn, newsletters

PROMPT: Act as a brand storyteller. Write a short post that shares a relatable behind-the-scenes moment or a small, human insight from my week. Please ensure the post: Feels honest, warm, and grounded — human but not heavy Shares a simple moment or observation that others can see themselves in Includes one gentle insight or reflection from the experience Builds trust and connection without oversharing or emotional intensity Fits easily into an automated or batched content system Tone: honest, warm, grounded, with steady emotional energy. Keep the post short, sincere, and connection focused. Behind-the-Scenes Moment: [Insert moment/insight] Audience: [Insert audience] Brand Voice (optional): [Insert brand voice]

Why It Works: Relatable, Short, Builds emotional connection.

Template 3: The Soft Invitation (No Pressure Promotion) Perfect for product launches, book promotions, service reminders, or inviting people to take the next step.

TEMPLATE: If you're ready for [desired result], I've created something that can help: [offer/book/service]. It's designed for [specific audience] who want [core benefit or outcome] – without feeling overwhelmed or adding more noise to their day. If it feels like the right next step for you, you can explore it here: [link or call to action]. Tone: Warm, calm, low-pressure, service-oriented Best for: Product launches Book promotions Service reminders Gentle invitations to take the next step Platforms: Instagram captions, Facebook posts, LinkedIn, newsletters, landing pages

PROMPT: Act as a relationship-based marketer. Write a soft, no-pressure promotional post for my [product/service/book]. Please ensure the post: Uses a warm, welcoming, and genuinely helpful tone Highlights one key benefit in a clear, simple way Gently invites the reader to learn more – without any hard-sell language Feels human, grounded, and respectful of busy, overwhelmed audiences Aligns with a relationship-first approach rather than urgency-driven marketing The final post should be: Short Supportive Easy to personalize Suitable for social media, newsletters, or community posts Offer: [Insert product/service/book] Audience: [Insert audience] Brand Voice (optional): [Insert brand voice] Key Benefit: [Insert benefit]

Why It Works: Never pushy. Focuses on benefit. Clear invitation. Emotionally safe for reader

Template 4: The 3-Part Email Simple, warm, and effective – perfect for newsletters, announcements, or reconnecting with your list.

TEMPLATE: Paragraph 1 – Relate (Human Connection) Here's something I've been thinking about lately: [insert insight]. It came up for me during [brief moment, challenge, or observation], and I have a feeling I'm not the only one who's felt this. Paragraph 2 – Teach (Offer a Simple Lesson) It reminded me that [insert lesson or truth] — a pattern I see often and something many of us navigate in real life. A small shift like [insert practical tip or perspective] can make things feel a little lighter. Paragraph 3 — Invite (Gentle Next Step) If you're working on this too, I created something that might support you: [offer/book/resource]. No pressure at all — just an option if it feels helpful for where you are right now. Tone: Warm, grounded, encouraging, and human Best for: Newsletters Announcements Nurture Emails Reconnecting with your list Soft promotional emails

PROMPT: Act as an email copywriter. Create a simple, warm 3-part email for my audience of [describe your audience]. Please structure the email as follows: 1. Relatable Opening Begin with a short, human moment, observation, or feeling that your audience can immediately connect with. 2. One Helpful Insight or Update Share one practical tip, supportive perspective, or meaningful update — something useful and easy to absorb. 3. Gentle Closing Invitation End with a soft, no-pressure invitation to explore[offer/book/resource], or to reply, reflect, or continue the conversation. Tone must be: Friendly Human Easy to read Supportive, not salesy Keep the entire email brief and scannable, suitable for busy readers. Audience: [Insert audience] Offer/Update: [Insert offer or context] Brand Voice (optional): [Insert brand voice]

Why It Works: Conversational. Balanced. Builds trust before offering

Template 5: The Repurposing Carousel (5 Slides) Turn any idea into a scroll-stopping carousel.

TEMPLATE: Slide 1 — HOOK (Stop the Scroll) Struggling with [specific problem]? Try this simple shift. Slide 2 — INSIGHT (Name the Truth) Here's the real reason this feels hard: [short explanation that feels human and relatable]. When you understand this, the challenge becomes much easier to navigate. Slide 3 — TEACH (Offer the Solution) Here's one small adjustment that helps: [practical tip]. Keep it simple — this works best when applied gently and consistently. Slide 4 — EXAMPLE (Show, Don't Just Tell) Try it like this: [example or scenario]. This gives people a clear way to picture the shift in action. Slide 5 — INVITE (Soft CTA) If you want more support with this, I created [offer/book/resource] to help. No pressure — just a tool for when you're ready. Tone: Warm, clear, educational, and low-pressure Best for: Instagram, LinkedIn, Facebook carousels, slide decks, or repurposing long-form content

Why it works: This structure teaches, supports, and gently invites without overwhelming — ideal for busy audiences and automated content workflows.

PROMPT: Act as a content repurposing strategist. Turn the following idea or short piece of content into a 5-slide social media carousel. Each slide must be clear, visually simple, and crafted to maintain scroll momentum. Please structure the carousel as follows: Slide 1 — Hook A short, compelling statement or question that immediately grabs attention. Slides 2–4 — Key Points Three concise, high-value points that build on the idea. Each should be: Easy to read Visually Scannable Focused on one message at a time Slide 5 — Gentle Call to Action A warm, no-pressure invitation (learn more, reflect, save, share, explore the book, etc.). Tone must remain supportive, clear, and aligned with my brand voice. Idea or Content: [Paste idea here] Audience: [Insert audience] Brand Voice (optional): [Insert brand voice] Offer or CTA (optional): [Insert offer/book/resource]

Template 6: The DM Follow-Up (Warm, Easy) For responding to comments, likes, or private messages — without sounding scripted.

TEMPLATE: Thank you so much for engaging — I really appreciate it. If you ever want support with **[topic]**, feel free to reach out. No pressure at all — I'm here to help whenever it's useful for you.

Why This Template Works: Warm, and unscripted: It feels personal even when reused. Zero pressure. Makes people feel safe to ask questions. Flexible: Works for comments, likes, DMs, saves, shares, or link clicks. Perfect for automation: Can be used by you or a VA without losing authenticity.

PROMPT: Act as a relationship-based marketer. Write a warm, human DM follow-up to someone who engaged with my recent post about [insert topic]. Please ensure the message: Feels conversational, friendly, and unscripted — like a genuine check-in Acknowledges their engagement (like, comment, share, save, etc.) Offers a simple, human point of connection related to the topic Contains no sales language or pressure Includes one light, optional invitation to continue the conversation if they're interested Tone: warm, grounded, supportive, and true to relationship-first marketing. Keep the message short, easy to personalise, and suitable for manual or automated sending. Topic: [Insert topic] Audience: [Insert audience] Brand Voice (optional): [Insert brand voice]

Template 7: The "Why I Made This" Post Ideal for books, digital products, and services.

TEMPLATE (Copy, Paste, Personalize): Thank you so much for engaging — I genuinely appreciate it. If you ever want support with [topic], feel free to reach out. No pressure at all — I'm here whenever it's helpful for you.

Why This Template Works: 1. Warm, and unscripted Feels personal even when used repeatedly, keeping the interaction authentic. 2. Zero-pressure by design Creates a safe, open space for people to ask questions without feeling sold to. 3. Highly flexible Works for comments, likes, shares, DMs, saves, story replies, or link clicks. 4. Perfect for automation and delegation A VA or social media manager can use this without compromising tone or trust.

PROMPT: Act as a brand storyteller. Write a short post explaining why I created my [book/product/service]. Please ensure the post clearly conveys: The problem it solves — in simple, relatable language The people it helps — who they are and what they struggle with the intention behind it — the deeper purpose or motivation A tone that is sincere, grounded, and relatable Zero sales pressure — the focus is connection, not conversion A gentle invitation to learn more at the end Keep the writing warm, human, and easy to read. Suitable for social media, newsletters, or websites. Offer: [Insert book/product/service] Audience: [Insert audience] Problem: [Insert problem] Brand Voice (optional): [Insert brand voice]

Why It Works: Heartfelt. Transparent. Connects intention with value

Template 8: The Weekly Recap Post Perfect for engagement and showcasing progress.

TEMPLATE (Copy, Paste, Personalize): This week, I focused on: • [Task or theme 1] • [Task or theme 2] • [Task or theme 3] What worked well: • [Insight, shift, or small win] What I'm taking into next week: • [One intention or simple next step] If you're working on similar goals, I hope this reminds you that progress doesn't have to be loud to be meaningful.

Why This Template Works; Encourages engagement through transparency and relatability. Makes your progress visible in a calm, structured way. Easy to customise weekly with minimal effort.

PROMPT: Act as a social engagement strategist. Write a short weekly recap post that highlights my key wins, insights, or actions from this week Please ensure the post includes: A simple, positive, reflective opening A concise list or summary of this week's wins, insights, or actions One practical takeaway my audience can apply or reflect on A light question or gentle CTA to encourage engagement (e.g., "What did this week teach you?") A tone that feels warm, grounded, and human — not overly polished or salesy Keep the structure clean, easy to read, and ready to post. This Week's Highlights: [Insert wins/insights/actions] Audience: [Insert audience] Brand Voice (optional): [Insert brand voice]

Template 9: The Book Teaser Template A simple way to tease or highlight your book.

TEMPLATE (Copy, Paste, Personalize): A little excerpt from my book today: "[insert short quote]" If this resonates, you'll really enjoy Chapter [number], where I dive deeper into [theme or benefit]. More to come — I can't wait to share this with you. Optional Add-On Lines (Use When You Want More Engagement): "Does this line speak to you too?" "This was one of my favorite sections to write." "I wrote this for anyone who's navigating [insert challenge]."

Why This Template Works: Short, polished, and easy to reuse. Showcases the book without pressure. Sparks curiosity by highlighting one meaningful excerpt.

PROMPT: Act as a book marketing strategist. Write a short teaser post that highlights one idea, insight, or theme from my book [insert book title]. Please ensure the post: Feels intriguing and accessible — easy for any reader to connect with Focuses on one key idea from the book Includes one clear benefit for the reader Maintains a warm, grounded, reader-focused tone Ends with a gentle invitation to stay tuned or learn more — absolutely no hard sell Feels social-media-ready and in order Book Title: [Insert book title] Theme/Insight: [Insert idea] Audience: [Insert audience] Brand Voice (optional): [Insert brand voice]

Why It Works: Quick. Enticing. Gives value through a quote

Template 10: The Launch Day Template Stress-free and easy to personalize.

TEMPLATE (Copy, Paste, Personalize): It's launch day — and I'm thrilled to finally share this with you. If you've been looking for [desired transformation or result], this book was written with you in mind. Thank you to everyone who has supported this journey. Your encouragement means more than you know, and I'm grateful you're here as this next chapter begins. If you'd like to explore the book, you can find it here: [link]. Optional Personalisation Add-Ons (Use as needed): "This project was built for busy marketers who want clarity without the overwhelm." "I wrote this during some of my busiest seasons — which is why it's designed to be simple and supportive." "If you're navigating [challenge], I hope this book feels like a breath of relief.

Why This Template Works

Creates excitement without hype Feels personal and grateful Offers a clear transformation without pressure Extremely easy to customize for any launch

PROMPT: Act as a launch copywriter. Create a simple, stress-free launch day post for my [book/product/service]. Please ensure the post includes: A warm, celebratory opening that feels human and grounded A clear line stating what's launching A short explanation of who it's for One clear benefit or outcome the offer provides A gentle, no-pressure invitation to check it out A tone that is warm, calm, encouraging, and easy to personalize The post should be easy to read, friendly, and suitable for social media or email. Offer: [Insert book/product/service] Audience: [Insert audience] Benefit: [Insert main benefit] Brand Voice (optional): [Insert brand voice]

These templates help you express your message with ease.

Closing Reflection

This chapter focused on more than saving time. It addressed mental space—reducing cognitive strain and creating systems that support creativity and sustainability. You are no longer relying on effort alone. You have established a foundation designed to support your pace and responsibilities without unnecessary friction.

Forward Into Chapter Nine

With your systems in place, the focus now shifts to sustaining momentum over time. Chapter Nine explores how to remain connected to your work during busy seasons, recognise natural creative cycles, and design expectations that prevent burnout. This chapter is about staying steady—not through pressure, but through a clear understanding of how you function best

Chapter Nine

Maintaining Momentum: How to Stay Consistent Without Burning Out

Section One

Consistency isn't difficult because people lack discipline — it's difficult because life is full. When your mind is stretched in ten different directions, even the best intentions can lose momentum — from overload. No marketer, no matter how disciplined, can produce steady content while running on empty. We need to develop a rhythm that supports us, a system that gives us freedom, not pressure. Consistency isn't about posting constantly. It's about staying connected to your work being present often enough that you don't lose momentum, repeating small actions — and being present for each move forwards. This is sustainable consistency — the kind that lasts.

The 3 Rules of Sustainable Consistency

These rules form the backbone of marketing rhythm for people with busy, unpredictable lives.

Rule 1 — Show Up Weekly, Not Daily: You do not need to create content every day. Weekly showing-up is enough to: maintain visibility, stay connected with your audience, build trust, move your business forward promote your book or offers gently. Daily posting is optional. Weekly posting is sustainable.

PROMPT: Act as a marketing strategist. Create one simple weekly visibility post for my audience of [describe your audience]. The post should maintain connection, build trust, and offer gentle value without pressure. Keep it short, clear, and effective for busy weeks when daily posting is unrealistic.

Rule 2 – Prepare When You Have Energy, Automate When You Don't: There will be days when inspiration flows – and days when you can barely think straight. Use high-energy days for: batching posts, recording videos, building ideas, writing longer-form content Use low-energy days for: scheduling, repurposing, soft engagement, responding to comments, posting templates you've already prepared

PROMPT: Act as my creative energy partner. Help me sort the ideas below into two clear categories: 1. Create on High-Energy Days Tasks that require creativity, original thinking, strategic decision-making, deeper focus or emotional bandwidth 2. Automate for Low-Energy Days. Tasks that can be, streamlined, templated, repurposed, scheduled, completed with minimal cognitive load. Please ensure the final output is simple, organized, and readable, aligned with a sustainable marketing rhythm, supportive for days when energy fluctuates, focused on reducing burnout and maintaining momentum Ideas to Sort: [Paste your ideas here] Audience: [Insert audience] Brand Voice (optional): [Insert brand voice]

Rule 3 – Keep a 'Momentum Folder' You Can Grab from Anytime. This folder is your safety net. Inside it you'll keep half-written posts, saved prompts, caption templates, repurposed pieces, ideas, quotes from your book, questions readers ask, FAQs, testimonials or comments, screenshots of messages On tired days, you don't create – you *pull* from the Momentum Folder. You're never starting from zero.

PROMPT: Act as my organization assistant. Organize the notes below into a Momentum Folder with clear categories and practical suggestions for how to use each piece. Please include: 1. Categories Sort the notes into helpful, strategic groups such as: High-Energy Creation, Low-Energy Automation, Repurposing Opportunities, Evergreen Ideas, Quick Wins, Drafts & Work-in-Progress (Feel free to add or refine categories based on the content.) 2. Usage Suggestions For each item, add a short note on how to use it, when to use it (high-energy / low-energy days), and what format(s) it could become (caption, reel, email, carousel, etc.).

3. A Simple Momentum Summary

PROMPT: Act as my clarity and organization partner. Write a short, supportive Momentum Summary that explains how to use this folder to stay consistent without burning out. Please ensure the summary: Feels calm, clear, structured, and supportive Emphasizes working with my energy, not against it Explains how this folder turns clarity into simple, sustainable action Reinforces that momentum comes from small, doable steps Is written in a warm, grounded tone suited for busy, real-life marketers The summary should be 1 short paragraph, easy to read, and actionable. Notes: [Paste your notes here] Audience: [Insert audience] Brand Voice (optional): [Insert brand voice] Use this as the guiding concept: "How to Maintain Momentum Without Burning Out: This is where clarity turns into behavior — gentle, doable actions you can rely on."

1. Set One Monthly Focus

Every month, choose one theme to guide your content. Not three. Not five. Just one. It becomes your North Star.

PROMPT: Act as my strategist. Based on my goals, suggest one monthly focus I can build my content around. Please include: A clear monthly theme aligned with my goals Why this focus matters for my audience and my long-term strategy How this focus supports consistency without increasing workload One sentence describing how the theme can guide weekly content (Teach, Connect, Invite, or your framework) The final output should feel strategic, sustainable, and easy to execute, helping me maintain momentum without burning out. My Goals: [Insert your goals] Audience: [Insert audience] Brand Voice (optional): [Insert brand voice]

2. Use AI as a Reset Button On days when you're overwhelmed, stuck, or blank, use this:

PROMPT: Act as my clarity coach. I'm feeling mentally crowded today and don't have the bandwidth for complex thinking. Help me create one simple, usable post based on this idea: [insert idea]. Please ensure the post: Feels kind, gentle, and easy Reduces decision-making Uses clear, supportive language Requires minimal energy for me to edit or publish Tone: soft, steady, and pressure-free — designed for low-capacity days when I need a reset rather than inspiration. Idea: [Insert idea] Audience: [Insert audience] Brand Voice (optional): [Insert brand voice]

3. Celebrate Micro-Wins

Momentum grows through recognition — not pressure. Examples of micro-wins: posting one caption, turning one idea into a draft, replying to one comment, scheduling two posts, finishing a single paragraph, pulling one idea from your Momentum Folder, Every win counts. Every small action compound over time.

PROMPT: Act as my progress coach. Here are my small wins this week: [list]. Reflect them back to me in a motivating, grounded, and encouraging way. Please ensure the reflection: Highlights the effort and intention behind the wins Shows how these actions contribute to long-term momentum Reinforces that small steps count and compound over time Uses a tone that is warm, supportive, and confidence-building Leaves me feeling capable, steady, and proud — not pressured Keep the response short, human, and emotionally steady. Weekly Wins: [Insert list] Audience: [Insert audience, if relevant] Brand Voice (optional): [Insert brand voice]

Consistency is simply staying in contact with your work, week after week, season after season — using AI as a support system. This creates, longevity, stability, visibility, confidence, lower stress, healthier emotions, meaningful growth.

This is the kind of consistency a busy marketers' life can maintain.

Section Two

The Emotional Side of Marketing: Staying Grounded Through Highs and Lows

If you've ever felt anxious before hitting "post," or discouraged when engagement is low, or overwhelmed by the pressure to "be consistent," you are not alone. Marketing isn't just strategy. It's emotional work. It asks you to be visible. To express yourself. To share ideas that matter, to stay steady in a world that moves fast. This section is about the emotional resilience that you need to sustain your marketing — without burnout, fear, or self-judgment.

1. Understanding Emotional Cycles in Marketing: Your marketing will move through cycles, just like the rest of your life: High energy □ highly creative You post confidently, ideas flow, momentum feels natural. Low energy □ quiet, scattered You feel blank, tired, or unsure what to say. Discouragement □ self-doubt You question whether anyone cares or if your message matters. Renewal □ clarity returns. A new idea sparks; inspiration suddenly reappears. These cycles are normal. Consistency doesn't mean being in your *high-energy* cycle all the time. Consistency means learning how to support yourself through all the cycles.

PROMPT: Act as my emotional clarity coach. Help me understand what phase I'm in right now and what kind of marketing action best matches that phase. Please: Identify my current emotional phase based on the notes I provide (e.g., high-energy, low-energy, rebuilding, reflective, overwhelmed, inspired). Explain the phase briefly in clear, supportive language. Suggest one aligned marketing action that suits this phase — something sustainable, realistic, and momentum-building rather than draining. Keep the guidance gentle, grounded, and practical, helping me work with my energy instead of against it. Current Notes or Feelings: [Insert feelings/notes here] Audience: [Insert audience] Brand Voice (optional): [Insert brand voice]

2. The Fear of Visibility (And How to Grow Through It): Visibility doesn't require volume — it requires presence. You don't need a big personality. What you need are small, reliable steps, steps that assist you with clarity, consistency, and authenticity. Visibility isn't about being "more." It's about being *you* in a way that is steady, sustainable, and aligned with the work you want to be known for.

PROMPT: Act as a gentle visibility coach. Help me turn this idea into a post that feels safe, warm, and true to who I am. Please ensure the post: Reflects my voice and personality Feels emotionally safe — no pressure, no intensity Maintains a tone that is warm, grounded, and authentic Shares the idea simply, without overexposure Supports visibility without triggering overwhelm Encourages connection rather than performance Keep the writing soft, human, and easy to publish — perfect for days when showing up feels tender but important. Idea: [Insert idea] Audience: [Insert audience] Brand Voice (optional): [Insert brand voice]

3.Low engagement can feel painfully personal: Even when it has nothing to do with you. When a post doesn't perform the way you hoped, your mind can start whispering unkind things: "My writing must not be good enough." "Maybe I should stop posting." But engagement is shaped by countless factors you cannot control algorithms resetting, time zone mismatches, holiday seasons, audience fatigue, posting timing, content saturation — even pure randomness. The key is to meet low-engagement days with understanding, not judgment. Listen to your energy. Adjust your pace. And trust that consistency is built over months of showing up with heart and presence, not in the performance of a single post.

PROMPT: Act as my grounding coach. I'm feeling discouraged by low engagement. Help me: Reframe this feeling in a calm, supportive, perspective-building way Understand why this reaction is normal and not a sign of failure Choose one small, doable next action that supports steady momentum without pressure Tone must be gentle, grounding, and reassuring, with clear practical direction suited for real-life marketers. Context: [Insert any details about the post or situation] Audience: [Insert audience] Brand Voice (optional): [Insert brand voice]

4. Returning to Your Work After Life Interrupts You: You: There will be weeks when work overload intensifies, when deadlines collide, when unexpected challenges demand your attention, or when stress, grief, or frustration quietly steal our focus. Your marketing rhythm should bend with you, not break under pressure. And when life pulls you away — as it inevitably will — you need a way to return without starting from zero.

This is where AI becomes a powerful support, not a shortcut. When you're ready to re-engage, AI can help you sort through what matters, rebuild your momentum, and remove the heaviness of restarting.

Whether you need a few days off or an unexpected pause, tools like ChatGPT make it possible to return to your commitments with confidence — and still meet your deadlines without the panic spiral.

PROMPT: Act as my re-entry guide. I've had time away from my marketing. Help me choose one gentle next step that will ease me back into flow — without pressure, guilt, or overwhelm. Please ensure the guidance: Acknowledges the normalcy of stepping back. Suggests one simple, low-resistance action to restart momentum. Feels soft, grounding, and supportive. Helps me reconnect with my audience and my message at a pace that honors where I am. Encourages sustainable consistency, not a surge of effort. Tone: warm, compassionate, steady, and suitable for marketers returning from a pause. Context: [Insert context about why you stepped away, if desired] Audience: [Insert audience] Brand Voice (optional): [Insert brand voice]

5. Protecting Your Energy Without Disconnecting from Your Audience: You can be present for your audience without sacrificing your wellbeing. You can engage meaningfully without overextending. You can respond with care without absorbing other people's energy or expectations. And learning to balance these is one of the most important parts of building a sustainable marketing rhythm.

Every small step you take in your work creates a ripple of momentum inside you. When you finish a caption, publish a post, assemble a campaign idea, or refine a single sentence, your brain releases a little surge of dopamine — the "you did it" chemical. These tiny wins matter. They energize you. They reinforce progress. They remind you that you can move forward, even on days when your energy is low or your confidence wavers. One of the best ways to protect your energy is to harvest and anchor those energizing moments.

When you complete a task, take 5 seconds to acknowledge it. When you receive a kind comment or message, pause to let it land. When a prompt works beautifully, smile and save it. When you make progress during a busy or emotional day, celebrate that resilience. These micro-celebrations refill your internal battery. They help you stay fueled rather than drained.

Protecting your energy isn't about pulling away from your audience — it's about showing up in ways that allow you to stay present without becoming depleted. When you move this way, your audience doesn't lose you — they receive the best of you: grounded, steady, and authentically engaged.

PROMPT: Act as my boundary coach. Write five warm, supportive replies I can use to engage with my audience without giving too much of my emotional energy. Please ensure each reply: Feels kind, appreciative, and human Creates connection without inviting emotional labor Uses short, simple language that is easy to paste or customize Maintains supportive boundaries while still nurturing engagement Works for comments, DMs, story replies, or post interactions Tone: warm, grounded, gentle, and emotionally sustainable. Context or Topic (optional): [Insert topic] Audience: [Insert audience] Brand Voice (optional): [Insert brand voice]

This helps you stay present in a balanced, grounded way

6. Letting AI Support, You Emotionally (Not Just Strategically): AI is not just a tool. It can also be a calming presence — a thinking partner, a clarity companion, and sometimes even the gentle voice that steadies you when your own thoughts feel tangled. It's easy to assume AI is only here to make tasks faster, but the truth is, it can also help you navigate the emotional weight of a busy, overloaded creative life.

Use AI in moments when your mind feels heavy or clouded. Use it when you're tired, unsure, muddled, or stretched thin. Use it when life has taken the wind out of your sails and you need clarity without pressure. AI is here to make your marketing life easier.

PROMPT: Act as my emotional support assistant. Help me turn this messy feeling into one simple, doable marketing action: [describe fe eling].Please ensure your response: Acknowledges and normalises the feeling in a gentle, grounding way Reflects back the emotional state with clarity Suggests one low-pressure action that matches my current capacity Keeps the step small, realistic, and momentum-building Maintains a tone that is steady, kind, and supportive, not pushy This action should help me move forward without increasing stress or overwhelm. Feeling: [Insert feeling] Audience: [Insert audience] Brand Voice (optional): [Insert brand voice]

This transforms emotional weight into clarity and movement

7. Anchoring Yourself in Purpose When Doubt Appears: Every marketer, creator, and author — no matter how skilled or experienced — faces moments when doubt creeps in. It doesn't matter whether you've published ten books or you're writing your first caption; the mind still whispers, "Is *this good enough*?" or "*Who am I to share this*?" Doubt is universal. But purpose is what brings you back to center.

Your purpose is the quiet thread running beneath everything you create. It's the reason you sit down to write even when you're tired, the reason you return to your work after setbacks, and the reason your message matters more than your fear. When doubt rises, purpose is the anchor that steadies you.

Here's a simple grounding practice you can return to when in doubt. Someone out there needs clarity, encouragement, or direction — and your work is part of their path.

Why they need your voice: Your perspective comes from real life, not theory. It carries weight and lives truth.

What your work gives them: A shortcut. A sense of calm. A system. A spark. A moment of relief.

How your experience supports theirs: Every challenge you've overcome, every tool you've learned, and every insight you've gained becomes a gift you can pass on. Sometimes the simplest reminder is the most powerful: Your voice matters because someone needs it. Your work matters because someone is waiting for it. Purpose doesn't remove doubt — but it gives you something stronger to walk with.

PROMPT: Act as my purpose mirror. Reflect back to me why my work matters and who it helps. Please include: A clear, grounded reminder of the purpose behind my work. A reflection of who benefits from what I create or share. One or two sentences that reconnect me to the impact I'm making. A tone that is encouraging, steady, and sincere — not dramatic, not salesy. A gentle closing reminder that purpose fuels momentum, even on low-energy days. Keep the reflection short, human, and emotionally supportive. Context or Work Description: [Insert details] Audience: [Insert audience] Brand Voice (optional): [Insert brand voice]

This reconnects you to the heart of your marketing — which is always people.

Marketing isn't just a workflow. It's an emotional journey — and one you no longer have to navigate alone. **N**ow we have engineered grounding prompts, emotional support prompts, visibility confidence tools; encouragement templates; re-entry routines, healthy boundaries, self-compassion frameworks. This helps us stay steady through the highs and lows of showing up, sharing our voice, and building something meaningful, feeling supported, and empowered — and we can move forward with confidence.

Section Three

A Sustainable 12-Month Rhythm: Planning Your Marketing Year Without Overwhelm

Most yearly marketing plans fail for one simple reason: They're built for people with endless energy, predictable routines, and daily creative output. Real marketers don't live like that. We certainly don't live like that. Real life has heavy months and light months, emotional highs and lows, busy seasons and slower seasons, health dips, and work demands, unexpected interruptions, waves of inspiration followed by silence.

That's why your year doesn't need a strict marketing plan. It needs a rhythm – a gentle roadmap you can follow without feeling trapped by it. Let's create a yearly rhythm that supports your reality and carries your marketing – and your message – with steady, sustainable energy.

Step 1 – Choose Your 4 Quarterly Themes: Each quarter has its own emotional tone and energetic capacity. Instead of over planning, choose *four simple themes* – one per quarter.

Examples:

Q1 – Fresh Starts & Visibility

Q2 – Growth & Education

Q3 – Community & Engagement

Q4 – Reflection & Renewal

PROMPT: Act as my yearly strategist. Based on my goals, suggest four quarterly themes I can build my content and marketing around. Please ensure each quarterly theme includes: A clear strategic focus aligned with my long-term goals Why this theme matters for growth, visibility, or momentum How the theme supports sustainable consistency (not hustle) The type of content that fits naturally with that quarter's energy (e.g., teaching, storytelling, launching, nurturing) The final output must feel: Structured, grounded, and achievable Supportive for both high-energy and low-energy seasons Designed to build momentum steadily across the year My Goals: [Insert goals] Audience: [Insert audience] Brand Voice (optional): [Insert brand voice]

This becomes the skeleton of your entire year.

Step 2 – Assign a Light Monthly Focus (12 Simple Anchors): Each month should have *one* focus. Not dozens. Not five campaigns. Just one.

Examples:

January – Clarity & Reconnection

February – Trust Building

March – Small Wins & Progress

April – Education & Value Posts

May – Audience Warm-Up

June – Mid-Year Reset

July – Storytelling & Behind-the-Scenes

August – Visibility Boost

September – Offer Prep

October – Soft Launching

November – Seasonal Support

December – Gratitude & Reflection

PROMPT: Act as my monthly planning partner. Using my quarterly themes, suggest one monthly focus for each month of the year. Please ensure each monthly focus: Aligns directly with its quarterly theme

Is clear, simple, and achievable Supports sustainable momentum — not overwhelm Points toward a natural content rhythm (Teach, Connect, Invite or similar) Provides direction without requiring complex planning The final plan should feel grounded, structured, and easy to follow, giving me a year of focused content creation with minimal decision fatigue. Quarterly Themes: [Insert themes] Audience: [Insert audience] Brand Voice (optional): [Insert brand voice]

This removes the mental load of constantly deciding what to post.

Step 3 – Build a Light Seasonal Plan (3–5 Seasons Per Year): Seasons bring natural marketing opportunities: New Year, Back-to-school, EOFY, Spring reset, Holiday season, Mother's Day. Business planning cycles You don't need to operate in all of them. Choose the ones that match your audience and your energy.

PROMPT: Act as a seasonal strategist. Based on my business and book goals, identify the 3–5 seasonal moments I should plan content around this year. For each seasonal moment, please include Why it's relevant to my goals or audience What type of content naturally fits this moment How the season supports visibility, engagement, or sales One gentle suggestion for how to incorporate my book/product/service if appropriate Keep the recommendations simple, strategic, and sustainable, supporting consistency without creating overwhelm. Business Goals: [Insert goals] Book Goals: [Insert goals] Audience: [Insert audience] Brand Voice (optional): [Insert brand voice]

This gives your year natural structure without over commitment

Step 4 – Allocate One "Core Content Piece" Per Month: This is your anchor piece: a newsletter, a blog, a carousel, a short video, a chapter excerpt, a story-based reflection This piece becomes: - 6 repurposed posts - an email - a caption - a teaser - a quote - a question prompt. One-piece fuels the month.

PROMPT: Act as my content planner. Suggest one cornerstone content idea for each month, based on my yearly themes. For each month, please include: A clear cornerstone idea that aligns with the theme A brief note on why this content matters strategically the format that fits best (e.g., blog, newsletter, video, carousel, teaching post) A short description that guides the angle or message Cornerstone content should: Be high impact but not overwhelming Provide material that can be easily repurposed throughout the month Support consistency and reduce decision fatigue Build momentum toward my book and business goals Keep the full plan streamlined, practical, and energy-friendly. Yearly Themes: [Insert yearly themes] Audience: [Insert audience] Brand Voice (optional): [Insert brand voice

Step 5 – Create Your "12-Month Momentum Folder: This is where you store: unused posts, repurposed pieces, ideas, book excerpts, promo messages, seasonal drafts, images and graphics, saved comments or feedback. This becomes your personal archive of consistent momentum – ready to pull from anytime your energy dips.

PROMPT: Act as my systems organizer. Create a simple, clear folder structure for my 12-month momentum plan. Please ensure the structure includes: A top-level annual folder that holds the full system Quarterly folders aligned with my themes Monthly subfolders that support easy planning and consistency Space for content drafts, repurposing assets, and reflections A layout that is lightweight, intuitive, and easy to maintain – even on low-energy days Tone: practical, calm, and supportive, focused on helping me stay organized without complexity. Yearly Themes (optional): [Insert themes] Audience: [Insert audience] Brand Voice (optional): [Insert brand voice]

Step 6 – Identify Your Rest Months (Yes, Really): Rest is not optional. Rest is part of sustainability. Choose 1–2 months per year where your marketing becomes: softer, lighter, automated, repurposed. Your audience won't disappear. They'll simply experience you in a calmer season.

PROMPT: Act as my wellbeing strategist. Based on my yearly structure, suggest which months are best suited as rest months and explain how I can maintain gentle, low-energy visibility during those periods. Please include: A short explanation of why each recommended rest month makes strategic sense The type of minimal, low-effort content that keeps me visible without draining energy Suggestions for repurposing, evergreen posts, or light engagement rather than fresh creation Guidance that supports sustainability, wellbeing, and emotional steadiness A tone that is calm, reassuring, and practical, emphasizing long-term momentum over hustle This should help me build a yearly rhythm that protects my capacity while keeping my brand present. Yearly Structure or Themes: [Insert themes] Audience: [Insert audience] Brand Voice (optional): [Insert brand voice]

Step 7 – Review and Reset Quarterly (Not Daily, Not Weekly): Quarterly reflection keeps you aligned without overwhelming you. Every 3 months, ask: What flowed? What felt heavy? What performed well? What drained me? What can be simplified?

PROMPT: Act as my quarterly review coach. Help me: Reflect on my last three months Identify themes, progress, wins, and learning moments Create three simple, sustainable priorities for the next quarter Please ensure your response: Highlights what worked, what supported momentum, and what felt draining Frames challenges with compassion and clarity Offers three grounded priorities that align with my yearly themes and energy Keeps the tone calm, encouraging, and practical, avoiding pressure or urgency Helps me build confidence and direction for the quarter ahead Reflection Notes: [Insert notes from the last 3 months] Yearly Themes: [Insert themes] Audience: [Insert audience] Brand Voice (optional): [Insert brand voice] This keeps your year supportive – not stressful. Putting It All Together: Your 12-Month Rhythm

This is not a rigid plan. It's a living rhythm you can return to any time life shifts – a structure that holds your marketing steady without demanding more than you can give. With this rhythm, you now have: a clear, doable 12-month plan, seasonal support built in, space for rest and recovery, a structure that adapts to your life, clarity without pressure, purpose without hustle, visibility without burnout. You are building your year from a place of power and clarity.

Closing Reflection

This chapter reframed consistency as alignment rather than endurance. You learned how to work with your energy, honour natural seasons and responsibilities, and maintain direction without force. Burnout is not a failure of commitment; it is a signal of misalignment.

When marketing is designed around rhythm rather than pressure, momentum becomes sustainable.

Forward Into Chapter Ten

Chapter Ten shifts your perspective from individual tasks to an integrated marketing ecosystem. You will explore how content, offers, email, and audience work together, how to develop long-term themes that strengthen identity, and how to build on what already works instead of starting over.

This chapter invites you to step back, see the larger structure you are building, and understand how your brand can evolve with clarity and confidence over time

Chapter Ten

Your Long-Term Marketing Ecosystem: Building a Brand That Supports You

Section One

Prompt engineering simplifies and enhances your marketing skills, and it gets easier when you build a long-term ecosystem instead of reacting week to week. An ecosystem is not a rigid plan. It's a supportive structure that grows with you, a container for your ideas, your audience, your message, and your offers.

An ecosystem is the part of your business that keeps you grounded, systemized and visible even when your energy dips.

The Four Pillars of a Long-Term Marketing Ecosystem

Your ecosystem does not need to be difficult. It needs to be repeatable and aligned with your life. There are four pillars that every sustainable marketing ecosystem includes:

Pillar 1 – Your Core Message

Your core message is the heartbeat of your brand—the foundation of everything you communicate. It defines what you stand for, what you teach, and the transformation you offer. When your message is sharp, you write faster, maintain consistency, speak with clarity, attract aligned audiences, and repurpose assets efficiently. A clear message becomes your North Star, guiding every campaign, caption, and conversation.

PROMPT: Act as a brand messaging strategist. Help me define or refine my core message — what I stand for, what I teach, and the transformation I help my audience achieve. Please: Identify the core belief or philosophy driving my work Clarify the value or insight I consistently bring to my audience Articulate the transformation or outcome my work supports Summaries everything in 2–3 clear, memorable sentences that can serve as the foundation for all future content, campaigns, and conversations Keep the tone confident, grounded, and authentic, aligned with my long-term brand identity Details About My Work: [Insert details] Audience: [Insert audience] Brand Voice (optional): [Insert brand voice]

Pillar 2 – Your Content Foundations

Strong foundations simplify execution: choose one primary social platform, one long-form content hub (such as a blog, email newsletter, or YouTube channel), one content format you can deliver consistently, and one repurposing workflow. With these systems in place, you eliminate constant reinvention and build a predictable marketing rhythm.

PROMPT: Act as a content strategist. Help me define my core content foundations by identifying: My main social platform – where I show up consistently One long-form content home (blog, email newsletter, YouTube, or similar) One content type I can create reliably and sustainably A simple repurposing routine that expands my reach without extra effort After identifying these four elements, summarize everything into a clear, practical, overwhelm-free plan I can follow long term. Please ensure the final plan: Supports sustainable visibility Works for both high-energy and low-energy seasons Reduces decision fatigue Builds a strong, cohesive brand presence over time Details About My Business: [Insert details] Audience: [Insert audience] Capacity/Energy Notes (optional): [Insert notes] Brand Voice (optional): [Insert brand voice]

Pillar 3 – Your Offers (Present or Future)

Your marketing ecosystem exists to support your offer—whether that offer is a book, a service, a digital product, a workshop, a lead magnet, a newsletter, or a community. You do not need a catalogue of offers to build a sustainable brand; one well-designed offer, supported by the right systems, is enough to establish authority and fuel growth.

PROMPT: Act as a marketing strategist. Help me clarify my main offer (current or future) by identifying: What the offer is Who it's for the problem it solves the transformation or outcome it helps people achieve Then, explain how my content ecosystem — social content, long-form content, and repurposing — can support this single offer effectively, without needing multiple products or services. Please ensure your guidance: Emphasizes clarity, focus, and long-term sustainability Shows how each part of the ecosystem builds awareness, trust, and momentum toward the offer Helps me avoid overwhelm by anchoring my marketing around one core pathway Uses a tone that is clear, confident, and supportive Offer Details: [Insert offer description] Audience: [Insert audience] Brand Voice (optional): [Insert brand voice]

Pillar 4 — Your Audience Pathways

This pillar answers a critical marketing question: How does a new audience member progress from first discovering you to engaging with you over the long term? Your ecosystem should guide people through four natural stages:, Discovery - social content, search visibility, collaborations, Connection - email, direct messages, personal storytelling, Trust-building - valuable education, consistent themes, proof of expertise, Invitation - offers, services, events, or books, When your pathway is designed intentionally, your marketing stops feeling sporadic and becomes a predictable journey that nurtures relationships and leads to meaningful action.

PROMPT: Help me map out a simple, intentional pathway that guides my audience from discovering my work to engaging with me long term. Please build the journey around these four stages: Discovery - how new people first find or notice my work Connection - how they begin engaging or resonating Trust-Building - how I deepen credibility, relationship, and value Invitation - how I invite them into my offer, book, or next step in a warm, ethical, non-pushy way Then, summarize the full pathway in a clear visual-style or bullet-point map I can use to align my marketing efforts across platforms and content types. Ensure the final map: Is simple, practical, and easy to follow Works for both high-energy and low-energy seasons Supports long-term consistency, not hustle Shows how my ecosystem builds momentum over time Business Details: [Insert details] Main Offer: [Insert offer] Audience: [Insert audience] Brand Voice (optional): [Insert brand voice]

A healthy ecosystem does not drain your energy—it strengthens and sustains you. It provides stability, predictability, and clarity; it nurtures creativity, reinforces your identity, creates space to breathe, builds confidence in your message, and connects you with the people who value your work. Clarity is the foundation of a sustainable system: with only four pillars, you can build long-lasting momentum. You now have the direction and prompt to begin designing a system that grows with your life, not one that competes with it.

Section Two

Designing Your Long-Term Content Themes (Your Marketing Identity)

When you commit to long-term content themes, you are giving your creativity structure. You are building a body of work. You are becoming recognizable. You are creating clarity for yourself and your audience.

Without themes, every post feels like starting from zero. But when you do have themes: ideas flow more easily, you stop second-guessing yourself, you know exactly what to create, you avoid content fatigue, you build trust faster, your audience understands your value, your marketing feels lighter, your message stays clear.

Themes turn marketing from guesswork into rhythm.

Step 1 – Choose 3–5 Content Themes You Can See Yourself Talking About All Year. Think of these themes as your content pillars — not trends, not quick ideas, but areas of real knowledge, lived experience, and ongoing curiosity. For most, marketers, themes often include things like: Your main value or message (Clarity, empowerment, progress, simplicity): Your expertise or offer (Prompting, writing, health, coaching, planning, business support): Your audience's needs (Confidence, organization, encouragement, focus, motivation): Your personal story (Life lessons, responsibilities, resilience, real-world wisdom) Behind-the-scenes of your work (Process, tools, insights, challenges, small wins): These become your brand identity — not what you sell, but what you stand for.

PROMPT: Act as a brand content strategist. Help me choose 3–5 core content themes I can speak about all year — based on: My values. My expertise. My audience's needs. My lived experience. The behind-the-scenes of my work. For each theme: Provide a 1–2 sentence summary/ Explain how this theme strengthens my long-term brand identity. Ensure it supports consistency, clarity, and sustainable content creation The final set of themes should feel: Authentic to who I am Relevant to what my audience needs Aligned with my long-term brand Easy to return to, even in low-energy seasons Strong enough to anchor every piece of marketing I create Details About My Work: [Insert details] Audience: [Insert audience] Values or Brand Voice (optional): [Insert values/voice]

Step 2 – Turn Each Theme into Repeatable Subtopics. A theme is broad. Subtopics make it usable. Let's say one of your themes is "Helping Busy People Find Clarity." Subtopics might include how to simplify a task: how to sort out your week: how to use prompts to reduce overwhelm: how to repurpose content: how to build confidence in your message: how to return to marketing after a break: how to choose gentle productivity tools: Suddenly, one theme becomes dozens of posts.

PROMPT: Act as a content development strategist. Take my core content themes and turn each one into a list of repeatable, practical subtopics I can use for ongoing posts, emails, videos, or campaigns. For each theme, please: Generate specific, usable subtopics that lead to clear, actionable content Include a variety of angles — educational, storytelling, behind-the-scenes, mindset, how-to, examples, and gentle invitations Ensure the subtopics are simple, sustainable, and evergreen Provide enough variation so each theme can generate dozens of ideas throughout the year Keep the tone and approach aligned with long-term brand identity and audience needs The result should feel like a renewable idea bank I can return to repeatedly. Core Content Themes: [Insert themes] Audience: [Insert audience] Brand Voice (optional): [Insert brand voice]

Step 3 – Map Your Themes to Your Audience's Emotional Journey. Your content shouldn't just teach. It should support. Think about your audience's emotional cycles: overwhelmed. confused, hopeful, discouraged, curious, ready to take action, good content themes address all of these states.

For example:

1. When they feel overwhelmed: share clarity prompts
2. When they feel discouraged: share small-win encouragement
3. When they feel hopeful: share tools and workflows
4. When they feel curious: share behind-the-scenes content
5. When they're ready to act share simple calls to action

Themes become stronger when they reflect both the practical and the emotional needs of your audience.

PROMPT: Act as a marketing psychologist and content strategist. Map my core content themes to my audience's emotional journey – including moments when they feel: Overwhelmed. Confused. Hopeful Discouraged' Curious. Ready to take action. For each emotional state: Identify which of my content themes best support that moment. Explain why that theme meets the emotional need Provide a short, practical guideline for how to craft content that fits that state Ensure the guidance is emotionally aware, sustainable, and aligned with long-term brand identity. Finally, create a simple, clear emotional-content alignment guide I can use throughout the year to craft content that meets my audience where they are – without adding pressure or complexity. Core Content Themes: [Insert themes] Audience: [Insert audience] Brand Voice (optional): [Insert brand voice]

Step 4 – Assign Themes to Months or Quarters (Lightly): This is not rigid planning. This is creating flow.

For example:

Q1 – Visibility & Confidence

Q2 – Systems & Structure

Q3 – Creativity & Expansion

Q4 – Reflection & Renewal

Or monthly:

January – Clarity

February – Connection

March – Consistency

April – Content Foundations

May – Storytelling

June – Visibility

July – Mindset

August – Efficiency

September – Offer Development

October – Launch Preparation

November – Audience Warmth

December – Gratitude & Reflection

When themes sit inside a year, everything feels easier.

PROMPT: Act as my long-term planning partner. Help me assign my core content themes across 12 months in a way that feels natural, sustainable, and aligned with my energy, my audience's needs, and my long-term brand identity. Please ensure: Each theme is placed in months where it naturally fits — based on seasonality, audience mindset, or strategic moments The distribution avoids overwhelm and creates an easy, supportive rhythm Themes repeat in a way that reinforces brand consistency without becoming repetitive The plan supports both high-energy and low-energy periods The final schedule is presented as a clear, simple monthly map I can follow throughout the year Tone: calm, strategic, and clarity-driven, focused on long-term momentum rather than short-term hustle. Core Content Themes: [Insert themes] Audience: [Insert audience] Seasonal or Business Notes (optional): [Insert details] Brand Voice (optional): [Insert brand voice]

Step 5 – Let Your Themes Become Your Signature Over time; your audience will begin to recognize you by the themes you return to — not the trends you chase. Your themes become the steady thread that people can rely on. This is how brands are built — not through constant reinvention or aggressive advertising, but through consistency of purpose. When you repeatedly return to your core themes, they become your signature. Your signature shapes your identity. Your identity becomes the ecosystem your business lives inside. And when you show up with that kind of consistency, your audience doesn't just remember you — they trust you.

This approach moves marketers out of daily content panic and into strategic, sustainable creativity. With these tools, you now have a dependable system—one that supports your work, protects your energy, and grows your brand steadily and intentionally over time.

Section Three

How Your Ecosystem Supports Your Offers and Future Growth

Many businesses approach each offer as a separate campaign requiring new strategies and bursts of energy. A long-term marketing ecosystem changes that dynamic. Instead of treating offers as isolated events, your message forms the foundation from which products, services, and future creations flow naturally. Your ecosystem becomes the fertile soil where ideas take root and grow, without requiring you to start from zero each time. It quietly holds your offers, strengthens your voice, and prepares your business for sustainable future growth.

1. Your Ecosystem Creates Natural Demand (Without Forcing It)

When your content themes are clear and consistent, your audience begins to understand who you are what you care about, what you help them with, what problems you solve, what value you bring. This builds trust, the foundation of all marketing. By the time you introduce an offer — whether it's a book, a downloadable resource, a session, a service, or a digital product — your audience is already warmed up. You don't need heavy sales tactics. You don't need aggressive funnels. You don't need to "convince" anyone. Your ecosystem has been preparing the ground all along.

PROMPT: Act as a marketing strategist. Review my core content themes and show me how they naturally build demand for my [book/product/service]. Please: Explain how each theme helps my audience understand: Who I am What I care about What problems I solve The unique value I bring Identify the emotional and logical bridges each theme creates between my everyday content and my offer Show how these themes position me as a trusted guide — not a salesperson Create a simple audience warm-up plan that: Feels natural Builds trust over time Prepares people to say "yes "Requires no heavy sales tactics, urgent funnels, or pressure The final output should feel clear, calm, strategic, and aligned with a long-term ecosystem, not a short-term campaign. Core Content Themes: [Insert themes] Offer: [Insert book/product/service] Audience: [Insert audience] Brand Voice (optional): [Insert brand voice]

2. Your Ecosystem Helps Your Offers Evolve Smoothly

When you've built strong content foundations, new ideas don't feel disruptive. They feel like the next right step. Your ecosystem gives room for: new products, revised services, seasonal offers, expanded versions of what you already teach, companion resources, future books, membership or community ideas. Your audience doesn't get confused because your themes stay steady. The *expression* evolves while the foundation remains familiar.

PROMPT: Act as a brand and offer-development strategist. Review my current content ecosystem and show me how I can introduce new or evolving offers — such as: Updated services, Seasonal promotions, Digital products, Companion resources, Future books...in a way that feels natural, aligned, and coherent for my audience. Please: Identify how my existing content themes already pave the way for new offers Explain how each theme provides continuity, familiarity, and trust Recommend ways to introduce new offers without confusing or overwhelming followers Highlight the narrative bridges that turn everyday content into effortless offer readiness Provide a simple, sustainable path for growth that keeps my brand consistent as my product suite evolves The final guidance should feel strategic, grounded, and easy to implement, supporting long-term expansion rather than sudden pivots. Current Content Ecosystem: [Insert details] Core Content Themes: [Insert themes] Potential or Upcoming Offers: [Insert offers] Audience: [Insert audience] Brand Voice (optional): [Insert brand voice]

3. Your Ecosystem Makes Selling Feel Natural (Not Stressful)

When your message, content, and audience pathways are aligned, several shifts occur: your audience expects your offers, your offers feel supportive rather than sales-driven, you stop worrying about "selling," your launches feel softer and clearer, and your marketing becomes part of an ongoing rhythm. This is especially important for marketers who juggle multiple responsibilities and cannot devote their entire schedule to constant promotion. Instead of thinking, "I need to sell now," you begin thinking, "My ecosystem is always warming people up." Selling becomes an invitation rather than an interruption.

PROMPT: Act as a marketing strategist who specializes in gentle, aligned selling. Review my message, content themes, and audience pathways. Show me how to present my offers in a way that feels natural, supportive, and pressure-free. Please include How my content ecosystem can warm up my audience continuously, instead of relying on high-pressure launch cycles How each content theme gently leads toward understanding the value of my offer The emotional journey my audience takes before they feel ready to say "yes" Guidance on presenting my offer as a simple invitation, not a push A short, sustainable weekly rhythm for soft, steady sales — ideal for time-poor marketers with shifting energy Ensure your tone is: Calm Confident, Warm, Supportive, Clearly aligned with long-term, ethical marketing My Message: [Insert message] Core Content Themes: [Insert themes] Audience Pathways: [Insert pathways or notes] Offer: [Insert book/product/service] Brand Voice (optional): [Insert brand voice]

4. Your Ecosystem Supports You During High-Energy and Low-Energy Seasons

There will be months when you feel: highly creative, deeply focused, ready to teach and share, excited about new ideas. And there will be months when you feel: tired, overstretched, distracted, emotionally heavy, unsure what to say. A strong ecosystem adapts to you.

During low-energy seasons, your themes, templates, and repurposing systems carry you.

During high-energy seasons, your ecosystem amplifies your best ideas. Either way, you stay present without burning out.

PROMPT: Act as a content sustainability strategist. Help me design a marketing ecosystem that supports both my high-energy and low-energy seasons. Please: Review my themes, templates, and repurposing routines Show me how they can carry me when I'm tired, overstretched, or experiencing decision fatigue Show me how they can amplify my creativity when I'm energized, inspired, and ready to build momentum Create a simple, practical plan that keeps me present and consistent throughout the year — without burning out Ensure the ecosystem adapts to fluctuating energy and real-life constraints Use a tone that is calm, steady, encouraging, and strategically grounded Your plan should include: Low-energy mode: simplified tasks, gentle posting, minimal effort actions, repurposing options High-energy mode: creation sprints, deeper content, expansion opportunities, batching workflows Sustainable baseline: a weekly rhythm I can maintain regardless of energy levels Core Content Themes: [Insert themes] Templates & Systems: [Insert templates or systems] Audience: [Insert audience] Brand Voice (optional): [Insert brand voice]

5. Your Ecosystem Becomes the Base for Future Books, Courses, or Signature Work

Your long-term themes don't just support your marketing — they become the blueprint for your future creations. Your ecosystem becomes: a research library, a draft repository, a testing ground, a clarity tool, a confidence builder. When you're ready to create your next big project — a workshop, a guide, a book, a course, a lead magnet — you won't be starting from scratch. You'll be building from a foundation of content that already resonates.

PROMPT: Act as a creative development strategist. Review my core content ecosystem and show me how it can become the foundation for future projects such as: Books, Courses, Workshops, Guides, Signature offers, Digital products. Please: Identify which themes, posts, or insights have the potential to evolve into long-form content or future offers Explain how my ecosystem can serve as a: Research library - for gathering ideas, patterns, and audience feedback Testing ground - for validating concepts before developing them Confidence-building tool - helping me see what resonates and what I naturally teach Recommend a simple, structured plan for turning existing content into assets for future projects Keep the guidance clear, practical, and aligned with long-term sustainability Ensure the tone is encouraging, strategic, and grounded, not overwhelming The final plan should help me see how my weekly content is laying the groundwork for my next body of work. Core Content Ecosystem: [Insert details] Core Content Themes: [Insert themes] Audience: [Insert audience] Brand Voice (optional): [Insert brand voice]

6. Your Ecosystem Builds Long-Term Trust — the Kind People Buy From

Trust isn't built through one post or one email. It's built through recurring themes, steady presence, relatable stories, helpful content, emotional connection, consistency over time. When people trust you, they don't need much convincing. Your offers feel like an extension of the relationship you've been building. A long-term ecosystem is how you create that trust — quietly, steadily, sustainably

PROMPT: Act as a trust-building marketing strategist. Review my current content themes, posting rhythm, and audience interactions. Show me how to strengthen long-term trust through: Steady presence, Relatable storytelling. Helpful insights, Emotional connection Then, create a simple, sustainable plan that helps my audience feel supported consistently — so that when I share an offer, it feels like a natural continuation of the relationship, not a sales push. Please ensure your recommendations: Emphasize relationship-based marketing, not metrics Reflect how trust grows through frequency, tone, vulnerability, clarity, and relevance Suggest ways to maintain trust even during low-energy seasons Explain how each content theme contributes to trust-building over time Use a tone that is steady, confident, warm, and aligned with long-term brand development Content Themes: [Insert themes] Posting Rhythm: [Insert rhythm] Audience Interactions: [Insert notes or examples] Offer: [Insert offer] Brand Voice (optional): [Insert brand voice]

Your marketing identity becomes a strategic advantage. You no longer need aggressive tactics to succeed. Your ecosystem enables you to evolve with confidence. Most importantly, you can now see that a long-term marketing ecosystem does not simply support your business—it supports you. Your energy. Your creativity. Your growth. Your future. This is a marketing system real people can sustain—not for months, but for years.

Closing Reflection

This chapter brought your marketing into focus as a connected whole. You moved from isolated posts to a cohesive system where content, offers, voice, and purpose work together. What you have built is not short-term momentum, but a foundation designed to support the next phase of your business. You are no longer reacting or assembling ideas at the last minute. You are shaping a brand that can move forward with clarity, confidence, and direction.

Forward into Chapter Eleven

Bringing It All Together

Chapter Eleven consolidates everything you have built into a single, workable strategy. You will learn how to align content, rhythm, and systems into a clear plan, design routines you can return to consistently, and use AI as a structured creative partner rather than a crutch.

This is where insight becomes execution—and where clarity turns into action.

Chapter Eleven

Bringing It All Together: Your AI-Supported Marketing Plan

Section One

Your marketing plan must be simple, flexible, supportive, energy-aware, AI-assisted, and practical. This chapter brings together everything you have learned. A plan that adapts to you, protects your energy, grows your authority steadily over time. A plan that makes you think, "Yes, I can do this." This is where strategy becomes routine. This is where your marketing life begins to work for you.

1. Start With Your Ecosystem, Not Your To-Do List

Your ecosystem — your themes, your core message, your audience pathway — is the foundation. Once that is set, your marketing becomes easier, because every piece of content has a purpose. Instead of asking "What should I post today?" You begin asking: "Which part of my ecosystem needs a little attention this week?" This small shift brings clarity and reduces overwhelm immediately.

PROMPT: Act as a marketing strategist. Using my core message, content themes, and audience pathway as the foundation, tell me which part of my ecosystem needs attention this week. Please: Identify the ecosystem area most in need of support or strengthening right now (e.g., Discovery, Connection, Trust-Building, Invitation, or Consistency) Recommend 1–2 simple, purpose-driven actions I can take this week that align with my long-term ecosystem Ensure these actions help me avoid random posting, reactive tasks, or scattered decision-making Keep the guidance clear, calm, supportive, and overwhelm-free Make the actions realistic for a time-poor marketer with fluctuating energy Your recommendations should help me stay connected to my long-term strategy while being gentle with my capacity. Core Message: [Insert core message] Content Themes: [Insert themes] Audience Pathway Notes: [Insert pathway details] Current Capacity (optional): [Insert energy level or constraints] Brand Voice (optional): [Insert brand voice]

2. Use AI as Your Thinking Partner

AI helps you: brainstorm ideas, clarify messages, polish your writing, repurpose content, simplify your workload, create drafts quickly, reduce decision fatigue but the real power of AI comes from using it consistently in your workflow — not randomly when you're stuck. Your plan should have AI built in; the same way a kitchen has its essential appliances.

PROMPT: Act as my marketing thinking partner. Help me use AI consistently in my workflow this week by guiding me through: Brainstorming ideas, Clarifying my message. Polishing my writing. Repurposing existing content. Simplifying my workload. Please recommend 3–5 specific places where AI can: Reduce decision fatigue. Speed up my process. Support momentum on both high- and low-energy days. Become a built-in, reliable part of my weekly workflow (not something I turn to only when stuck). Ensure your suggestions: Align with my core message, content themes, and ecosystem Are simple, practical, and easy to implement immediately Support long-term sustainability, not short-term hustle Keep the tone calm, strategic, and supportive Current Projects or Content: [Insert details] Core Message: [Insert core message] Content Themes: [Insert themes] Audience Pathway: [Insert pathway notes] Brand Voice (optional): [Insert brand *voice*

3. Build Your Plan Around What You Can Sustain – Not What You Wish You Could Sustain

A sustainable plan is built around your real schedule, real capacity, real responsibilities, real energy, and real life. If your creative rhythm fluctuates, your plan should flex with you.

PROMPT: Act as a sustainable marketing strategist. Help me design a marketing plan based on my real capacity — not the ideal version of myself. Please: Consider my actual schedule, energy levels, responsibilities, health, and workload Recommend a simple, flexible plan I can maintain consistently without burnout Priorities quality over quantity, even if that means: Posting once a week Running one campaign per quarter Using repurposed content Show me how to create momentum through low-friction, sustainable actions Keep the guidance realistic, calm, and supportive, emphasizing long-term health and consistency Ensure the plan aligns with my core message, themes, and offer pathway Your final plan should feel like a gentle structure I can rely on — not another overwhelming to-do list. My True Capacity: [Insert notes] Schedule Constraints: [Insert details] Energy Patterns: [Insert details] Core Message: [Insert] Content Themes: [Insert] Audience Pathway: [Insert] Brand Voice (optional): [Insert voice]

4. Choose Your Monthly Anchor and Let Everything Flow from It

Your monthly anchor is one more piece of content that everything else repurposes from: a newsletter, a blog, a carousel, a video, a story sequence. This anchor becomes your steady through-line, even during low-energy phases. From this anchor, AI can quickly generate captions, emails, reels scripts, LinkedIn posts, DM responses, carousel outlines, teaser posts. Your plan develops into a layered system, not a scatter of tasks.

PROMPT: Act as a content systems strategist. Help me choose one monthly anchor piece of content based on my strengths, energy, and realistic capacity. Options include: a newsletter, blog, carousel, video, or story sequence. Then: Explain why this anchor format fits my strengths and capacity. Show me how to repurpose that single anchor into a full ecosystem of content, including: Social captions Emails Reel or video scripts LinkedIn posts DM replies Carousel outlines Teaser or promo posts Create a simple, layered repurposing plan that ensures my entire month flows from one core piece — instead of scattered, disconnected tasks. Keep the plan efficient, sustainable, and aligned with my brand voice and message. My strengths: [Insert] My capacity: [Insert] Preferred content types: [Insert] Core message/theme for this month: [Insert] Brand voice (optional): [Insert]

5. Integrate Your Seasonal Campaigns Without Stressing Your Schedule

Your sustainable plan should include: 3–5 light seasonal campaigns, quarterly moments of intentional visibility, gentle warm-up phases, simple offers or lead magnets. These are not meant to disrupt your plan. They enhance it. Some seasons you'll be hands-on. Other seasons, your marketing will run softly in the background.

PROMPT: Act as a seasonal marketing strategist. Help me plan 3–5 light, sustainable seasonal campaigns for the year, based on major holidays, industry rhythms, and my own business milestones. Please include Campaign Themes: Identify the most natural seasonal themes or moments for my brand and audience. Purpose & Fit: Explain how each campaign fits within my existing content ecosystem and monthly plan without disrupting my workload. Warm-Up Phases: Outline simple visibility-building and connection-building steps that gently prepare my audience ahead of each seasonal moment. Offer Ideas: Suggest light, doable offer concepts — such as checklists, small digital products, workshops, promotions, book highlights, or nurturing campaigns. Execution Plan: Provide a practical, ultra-simple plan for each campaign that aligns with the capacity of a time-poor marketer (e.g., 1–2 posts/week, light email flow, easy repurposing). Energy-Aware Design: Ensure campaigns enhance momentum and ease, rather than overwhelm, by showing where they fit in my existing monthly structure. Tone: strategic, supportive, calm, and designed for sustainability over hustle. My Business Milestones: [Insert] Audience: [Insert] Seasonal Anchors or Holidays I Want to Recognize (optional): [Insert] Brand Voice (optional): [Insert]

6. Build Your Review Ritual into Your Plan (Not as an Afterthought)

Your weekly and monthly reviews aren't a bonus — they are the emotional anchor of your system. They help you: see your progress, adjust your strategy, honor your energy, refine what's working, release what's not, stay aligned with your message. Your plan stays flexible because reflection is built into it.

PROMPT: Act as a supportive marketing mentor. Help me build simple weekly and monthly review rituals into my marketing plan — not as an afterthought, but as an essential part of my workflow. Please guide me through: Reflection: Help me reflect on my progress with clarity and kindness — what worked, what felt aligned, and where I showed up even in small ways. Strategy Adjustment: Show me what to refine, amplify, reduce, or release based on what I learned this week or month. Energy Awareness: Help me honour my real capacity by identifying where my energy was supported or strained and adjust my approach accordingly. Alignment Check: Ensure my content, message, and actions remain aligned with my core themes, core message, and season of life. Gentle Next Steps: Recommend one simple next step for the week and one calm, sustainable focus for the month — designed to keep me moving without pressure or overwhelm. Create a review ritual that is: warm, grounding, sustainable, human, and easy to maintain — even for a time-poor marketer. This Week's Activity: [Insert] This Month's Activity: [Insert] Energy Levels: [Insert] Audience: [Insert] Brand Voice (optional): [Insert]

7. Keep It Light, Keep It Yours

You will evolve. Your audience will evolve. The goal is not quantity. the goal is quality. The goal is continuity. You need to do what matters and do it, consistently

PROMPT: Act as a compassionate marketing strategist. Help me create a marketing plan that feels light, human, and truly mine — something flexible and adaptable that evolves with my energy, my life, and my audience. Please include Foundational Approach: Outline a simple, sustainable marketing structure that values continuity over perfection and supports long-term ease. Flexibility Built In: Show me how this plan can shift with my energy levels, schedule, and seasons of life — without guilt or pressure. Priorities That Matter: Identify the 1–3 elements that matter most for me to stay consistent, connected, and visible in a way that feels natural. Gentle Rhythm: Recommend a soft weekly or monthly rhythm that supports consistent presence without rigidity or overwhelm.

You now see that everything you've learned has a place, a supportive structure that holds your marketing steady and protects your energy. This is the kind of marketing plan you can follow — not for weeks, but for years.

Section Two

Your One-Page Marketing Plan (AI-Assisted)

A marketing plan doesn't need to be long. It doesn't need to be complicated. In fact, the best marketing plans — the ones marketers actually follow — fit on one page. One page brings clarity. One page removes overwhelm. One page keeps you focused. One page gives you something you can revisit each week without dread. And when that one page is supported by AI, it becomes even easier to build, maintain, and adapt over time. Consider this your blueprint — the plan that supports your life instead of competing with it.

The 5 Elements of a One-Page Marketing Plan

Your one-page plan should include only what matters most:

1. Your Core Message: The heartbeat of your brand — the sentence that grounds every piece of content you create. It answers three questions: Who do you help? What problem do you solve? What outcome do you create?

Example: help busy marketers and everyday creators find clarity, confidence, and sustainable momentum using simple AI-supported workflows.

PROMPT: Act as a brand messaging strategist. Help me craft one clear core message that grounds my entire brand. This message must answer three essential questions: Who do I help? What problem do I solve?

What outcome do I create? Write this as a single, strong, concise sentence that I can use across my website, social media, bios, email signatures, and campaigns. Please also ensure the message: Is easy for my audience to remember and repeat Reflects my tone and positioning Creates immediate clarity about the value I offer Aligns with my long-term marketing ecosystem My audience: [Insert] The problem: [Insert] The outcome: [Insert] Brand voice (optional): [Insert] Tone: confident, clear, grounded, and professional.

2. Your 3–5 Content Themes: These are the pillars of your marketing identity — the topics you return to regularly. Your themes ensure you never face a blank page, because your content always has direction.

Example Themes:

Clarity & mindset: AI-supported workflows: Practical marketing tips: Gentle productivity: Stories from real life

PROMPT: Act as a content strategist. Help me identify 3–5 core content themes that will become the pillars of my marketing identity. These themes should: Provide clear direction for my content Prevent blank-page moments. Anchor my brand message Give my audience a consistent, recognisable throughline. Support sustainable, long-term marketing without guesswork. For each theme: Name the theme clearly. Summarize it in 1–2 concise sentences. Explain how it reinforces my long-term brand message and audience connection. Show why this theme deserves ongoing attention throughout the year. Tone: strategic, clear, grounded, and supportive. My brand message: [Insert] My audience: [Insert] My expertise: [Insert] Long-term goals (optional): [Insert] Brand voice (optional): [Insert]

3. Your Monthly Anchor Content: This is your steady output — the one thing you commit to consistently. It might be a blog, a newsletter, a carousel, a video, a weekly tip, a story sequence. Everything else you post can be repurposed from this anchor. This alone saves hours of time and energy each month.

PROMPT: Act as a content systems strategist. Help me choose one monthly anchor piece of content — such as a blog, newsletter, carousel, video, weekly tip, or story sequence — based on my strengths, capacity, and audience needs. Then: Explain why this anchor format is the best fit for my current season, energy, and goals. Show me how to repurpose this single anchor into multiple smaller posts or assets (e.g., captions, reels, LinkedIn posts, email snippets, carousel slides, value posts, teaser messages). Provide a simple, repeatable repurposing plan I can follow every month, even when my time or energy is limited. Ensure the plan supports: consistency, clarity, efficiency, and reduced decision fatigue. Tone: calm, strategic, supportive, and focused on sustainability. My strengths: [Insert] My capacity: [Insert] My audience: [Insert] Preferred content formats (optional): [Insert] Brand voice (optional): [Insert]

4. Your Seasonal Campaigns (3–5 per year): These are your higher-energy seasons — the moments where you share an offer, lead magnet, or book with a little extra intention.

Examples:

New Year clarity campaign: Mid-year momentum reset: Back-to-school support: Holiday gift guide or book promo: AI awareness month: Spring cleans your business systems: These campaigns fit gently around your monthly rhythm.

PROMPT: Act as a seasonal marketing strategist. Help me plan 3–5 simple, sustainable seasonal campaigns for the year. Identify which offers, lead magnets, or promotions I should highlight during these higher-energy seasons, and outline a light, realistic campaign approach that fits my true capacity as a busy marketer. Please include Seasonal Opportunities: Identify the natural seasonal moments that make sense for my business, industry, or audience. Offer Alignment: Recommend which offers, lead magnets, or promotions should be spotlighted during each seasonal moment — and explain why they are a natural fit. Campaign Simplicity: Outline a calm, manageable approach for each campaign, including a light posting rhythm (e.g., 1–2 posts per week), one simple email, and easy repurposing steps. Capacity Fit: Ensure each campaign reflects my real energy levels, schedule, and responsibilities — not an idealized version of me. Intentional Strategy: Show me how these campaigns can integrate smoothly with my existing monthly content plan without causing overwhelm or disruption. Tone: calm, strategic, supportive, and grounded in sustainability. My audience: [Insert] My offers/lead magnets: [Insert] My yearly goals: [Insert] My capacity: [Insert] Brand voice (optional): [Insert]

5. Your Review routine (Weekly + Monthly): Your plan only works when reflection is built in., Your review habits keep your marketing grounded: What worked? What flowed? What drained me? What feels aligned? What should I adjust?

PROMPT: Act as a supportive marketing mentor. Help me design a simple weekly and monthly review routine that keeps my marketing grounded, aligned, and sustainable. Guide me through reflecting on: What worked. What felt natural or flowed easily. What drained me or felt forced. What felt aligned with my message and audience. What needs refining, simplifying, or letting go Then: Turn these reflections into a clear, repeatable review process I can use every week and every month. Keep the system calm, practical, and overwhelm-free. Ensure the ritual helps me improve steadily without pressure, shame, or perfectionism. Recommend one aligned action for the upcoming week and one gentle focus for the next month. Tone: warm, grounded, reflective, and supportive. This Week's Activity: [Insert] This Month's Activity: [Insert] Energy Levels: [Insert] Audience: [Insert] Brand Voice (optional): [Insert]

Your One-Page Plan (Put Together) When all five elements come together on one page, it looks something like this: Your One-Page Marketing Plan (Template)

1. Core Message: Your one-sentence clarity statement.

2. Content Themes: Theme 1. Theme 2. Theme 3. Theme 4. Theme 5 (optional)

3. Monthly Anchor Content: One main piece of content you commit to creating.

4. Seasonal Campaigns: Campaign 1. Campaign 2. Campaign 3. Campaign 4 (optional)

5. Review Order: Your weekly and monthly check-in questions.

This is your marketing system — simple, clear, supportive. No noise. No pressure. Just a quiet, confident plan that holds you through the year.

How AI Helps You Maintain This Plan Effortlessly

Once your one-page plan is created, AI becomes the engine that keeps it alive. AI can: generate ideas, refine your message, repurpose anchor content, draft campaign copy, create templates. help you review your progress. adjust your plan as your life changes, support you when energy is low, expand on what's already working. Your plan stays simple, but your support becomes powerful.

PROMPT: Act as my ongoing AI marketing partner. Using my one-page marketing plan as the foundation, help me maintain and evolve it effortlessly throughout the year. Support me by: Generating new ideas that align with my core message, brand themes, and audience needs. Refining my message as my clarity deepens or my positioning evolves. Repurposing my anchor content into smaller posts, emails, captions, carousels, reels, or conversations that save me time and energy. Drafting campaign copy for simple launches, seasonal highlights, or nurturing sequences. Creating reusable templates that speed up my weekly workflow. Reviewing my progress with calm, kind clarity — identifying what's working and what needs simplifying. Adjusting the plan gently when my life, responsibilities, or capacity shifts. Supporting me during low-energy weeks by identifying the easiest, lowest-effort actions that still keep me visible. Expanding what's already working so my momentum builds naturally without hustle. Then provide: A clear set of steps for how AI can keep my marketing plan simple, sustainable, and powerful each week and each month. Suggestions for how I can integrate this partnership seamlessly into my real-life routine. Tone: strategic, calm, supportive, and human centered. My one-page marketing plan: [Insert or paste] My capacity: [Insert] My strengths: [Insert] My audience: [Insert] Brand voice (optional): [Insert]

This makes your plan a living system – not a forgotten document.

Section Three

What Your First 90 Days Can Look Like

Think of these 90 days as your onboarding into a calmer, more confident marketing life. A season where you learn to rely on your ecosystem, where AI becomes part of your rhythm, and where consistency grows naturally – without force. Here's what your first three months can look like when you do this in a sustainable, centered way.

Month 1 – Build Your Foundations (Clarity & Confidence): The first month is about grounding yourself – not perfecting anything. Your goal is simple: Create clarity + start small. Focus on: Your Core Message: Write one sentence that explains who you help and how. Your Content Themes: Choose 3–5 themes that feel natural and sustainable.

Your Monthly Anchor: Decide on the one type of content you will produce each month. One Gentle Weekly Touchpoint: This could be: one post, one short email, one repurposed idea, one story sequence. The goal is presence, and performance will follow. Begin Using AI Consistently: Use AI for: brainstorming, cleaning up your writing, generating captions, simplifying ideas

PROMPT: Act as a marketing clarity coach. Help me build my Month 1 foundations by guiding me to create simple clarity around: My core message My audience My content direction Then provide 3–5 small, confidence-building actions I can take this month to ground myself – without aiming for perfection or overwhelm. Please ensure the guidance is: Beginner-friendly Steady and supportive Focused on clarity, not volume Designed to help me start strong with gentle momentum

Tone: warm, grounding, simple, and confidence-building. My current understanding of my message: [Insert] My audience: [Insert] My content platform(s): [Insert] My energy/capacity for Month 1: [Insert] Brand voice (optional): [Insert]

At the end of Month 1, you'll feel: lighter, clearer, more grounded, and far more capable than when you began.

Month 2 – Build Your Flow (Consistency & Ease): This is the month where your plan begins to feel real – and easier. Your focus: Practice your rhythm + trust the process. Create Your Monthly Anchor Content: One blog, one carousel, one email, or one video – whatever you chose.

Repurpose It Into 4–6 Smaller Pieces: AI makes this simple. This becomes your weekly content.

Explore One Seasonal Campaign: A simple, light campaign – not a launch. Something like a New Year clarity message, a mid-year reset, a holiday tip series. a "behind the scenes" storytelling week Strengthen One Audience Pathway: Choose one: Instagram, LinkedIn, Email, Blog traffic, Facebook group Pinterest.

You're not trying to grow fast – you're learning how your audience responds.

PROMPT: Act as my marketing flow strategist. Help me implement Month 2 of my plan by guiding me through four core steps: Create one monthly anchor content piece Support me in choosing and producing a single anchor format – blog, carousel, email, or video – based on my capacity and strengths. Repurpose that anchor into 4–6 smaller weekly posts using AI Show me how to break the anchor into digestible content pieces that build consistency without extra effort. Explore one light seasonal campaign idea that fits this month Suggest a simple, doable seasonal theme or moment I can acknowledge or leverage – without adding pressure or workload. Strengthen one audience pathway Help me choose one platform to nurture (Instagram, LinkedIn, email list, blog traffic, Pinterest) and outline a calm, steady action that builds long-term trust and visibility. Finally: Create a simple, steady monthly workflow I can repeat. Keep everything focused on practicing consistency, building confidence, and trusting the process — not hustling or overproducing. Tone: calm, structured, supportive, and pressure-free. My Month 1 foundation: [Insert] My strengths: [Insert] My capacity for Month 2: [Insert] Preferred anchor format: [Insert] Audience pathway to strengthen: [Insert] Brand voice (optional): [Insert]

At the end of Month 2, you'll feel: more confident, more refined, and more connected to your audience.

Month 3 — Build Your Momentum (Visibility & Gentle Expansion): By Month 3, your ecosystem has roots. You're not guessing anymore. You're not starting from scratch. Now your focus becomes:

Expand gently and refine what works. Optimize Your Anchor Content: Begin improving what worked well last month — AI makes this effortless. Strengthen Your Repurposing Routine: Turn 1 piece □ 10 pieces. Examples: quotes, carousels, reels scripts newsletter intros, email tips, micro stories, Q&A responses

Introduce One Offer or Lead Magnet: Nothing big. Nothing stressful. Something supportive, like a checklist, a mini guide, a template, a free chapter, a simple tool, a short session. Your ecosystem warms your audience — your offer gives them a next step. Review Your First 90 Days: Ask: What felt easy? What drained me? What surprised me? What performed well? What do I want more of? What do I want less of?

PROMPT: Act as my marketing momentum strategist. Help me implement Month 3 of my plan by guiding me through four core components: 1. Optimise last month's anchor content using AI Show me how to refine, expand, strengthen, or deepen my Month 2 anchor content so it continues working for me. This may include improving clarity, tightening structure, enhancing storytelling, or elevating value. 2. Strengthen my repurposing routine Help me turn one core content piece into 10 smaller pieces, such as: Quotes, Carousels Reel scripts, Newsletter intros. Email tips, Micro stories, Audience Q&A responses, social captions, LinkedIn reflections, short teaching posts. The goal is to build a repeatable repurposing habit that increases visibility without increasing workload. 3. Introduce one simple offer or lead magnet Recommend a supportive, low-pressure offer aligned with my message and capacity — such as a checklist, guide, mini-training, sample chapter, resource list, or soft promotion of an existing offer. Show me how to naturally weave it into my content without feeling salesy. 4. Review my first 90 days of marketing Guide me through a reflective review using prompts such as: What felt easy? What drained me? What surprised me? What performed well? What do I want more of? What do I want less of? Use these reflections to shape my next steps and help me refine what's working. Final Output Create a clear, confidence-building Month 3 plan that gently expands my visibility and strengthens my marketing rhythm — without overwhelm or pressure. Tone: steady, supportive, practical, and aligned with sustainable momentum. My Month 1 and Month 2 notes: [Insert] My capacity for Month 3: [Insert] My core content piece for repurposing: [Insert] Potential offer/lead magnet ideas: [Insert] Brand voice (optional): [Insert] At the end of Month 3, you'll feel: steady, proud, and genuinely capable of maintaining your system long-term.

What 90 Days of Gentle, Realistic Marketing Gives You:

Clarity, direction, a sustainable rhythm, a reliable content system, a relationship with your audience, a grounded connection to your work, confidence in your ability to create consistently a marketing plan that doesn't crumble when life gets busy.

Your first 90 days are not a test. They are a beginning. And beginnings deserve grace.

Closing Reflection

This chapter brought your work into full alignment. You learned how to integrate content, tools, and rhythm into a cohesive system, simplify your marketing into repeatable steps, and design a plan that holds steady even during your busiest weeks. AI now functions as a partner in your process—not a replacement for your judgment or voice.

You are no longer facing a blank page or a fragmented calendar. You are working from a roadmap that respects your time, energy, and direction.

Forward into the Closing Chapter

You have done the work, built systems you can trust, found your rhythm, clarified your message, and shaped a marketing practice that supports you.

The Closing Chapter is a moment of reflection—not to add more, but to recognise what you have built and how far you have come. Carry forward your clarity, momentum, and lived experience. You are no longer searching for a way to make marketing work. You have created one.

Chapter Twelve

The Closing Chapter: Final Encouragement for the Long Game

Reaching the final chapter of a book like this is not a passive achievement. It signals something far more meaningful than completion. It tells us that you stayed engaged. You followed the throughline. You invested attention, time, and intent—despite competing demands, and the reality of a full professional and personal life.

This book was never designed to be skimmed, bookmarked, or consumed in fragments without action. It was designed to meet you in the real conditions of modern marketing—where interruptions are constant, expectations are high, and time is always in short supply. The fact that you are here means you did more than collect ideas. You made space for clarity. You paused long enough to reassess how you work, not just what you produce. That alone places you ahead of most.

You have not just learned how to use and engineer prompts. You have learned how to think in systems. You have not just explored workflows. You have learned how to reduce friction and decision fatigue.

You have improved your output. You have built a structure that supports consistency without strain. This is what separates short-term productivity from long-term sustainability.

You now have:

1. A repeatable way to move from idea to execution

2. A framework for clarity when energy is low

3. A process for momentum that does not rely on motivation

4. A system that works across platforms, formats, and seasons

Real marketing happens alongside meetings, family responsibilities, health fluctuations, client demands, and the unpredictable rhythm of everyday life. This book was written with that reality in mind—and your progress through it confirms that you are capable of working effectively within constraints, The goal was never to turn you into a content machine. The goal was to give you **control**.

Control over: How you start. What you prioritise. When you pause. How you return That ability to return, without confusion, without rebuilding from zero—is one of the most undervalued skills in modern marketing. You now have it. Before you move forward, anchor these principles. They are not motivational slogans. They are operational truths.

1. You Are Not Behind; There is no universal timeline for relevance, visibility, or success. Marketing careers do not progress in straight lines, and brands are not built on schedules borrowed from others. Every experience you bring—industry shifts, career pivots, life responsibilities, constraints—has sharpened your judgment. What once felt like delay has, in fact, refined your perspective. You are not catching up. You are arriving informed.

2. Effectiveness Does Not Require Noise; Visibility does not require constant broadcasting. Authority does not require urgency. Influence does not require exhaustion. The systems you have built allow you to communicate clearly, consistently, and confidently—without shouting. Quiet marketing, when well-structured, is often more persuasive than relentless activity. Impact is created through relevance, not volume.

3. Consistency Is a Pattern; Consistency is not about daily output. It is about predictable return. You have learned how to: Pause without losing momentum. Resume without friction. Maintain coherence across time That is what sustainable consistency looks like in practice.

4. Small Wins Compound Faster Than You Expect; Marketing progress is cumulative. One prompt refined. One workflow reused. One message clarified. One asset repurposed. These actions may feel modest in isolation, but together they form a compounding system. The work you do now reduces the effort required later. That is leverage.

5. Your Voice Is an Asset, not a Variable; Your perspective is not something to minimise or neutralise. It is what differentiates your work in increasingly automated environments. AI did not replace your voice in this process. It amplified it. You remain the strategist. The decision-maker. The authority.

6. AI Is a Support System, not a Substitute; You have learned how to lead AI. When used well, AI removes cognitive load, accelerates execution, and reinforces clarity. When used poorly, it creates noise. You now know the difference.

The Journey Ahead Is Not Linear—and That Is an Advantage. From this point forward, your marketing journey will move in cycles: Periods of momentum. Periods of maintenance. Periods of reflection. Periods of renewal None of these phases represent failure or regression. They are structural components of long-term work. Quiet seasons deepen strategy. Busy seasons test systems. Slow periods refine clarity.

Because you have built an ecosystem—not a collection of one-off tactics—you can move through these cycles without losing coherence or confidence. This is what it means to market for the long game. A Sustainable Definition of Growth. Growth does not always look like expansion. Sometimes it looks like: Fewer decisions. Faster starts. Cleaner execution. Reduced stress. These are not compromises. They are markers of maturity.

The system you have built allows you to scale *when* appropriate and stabilise when necessary. That adaptability is a strategic advantage most marketers do not possess. A Practical Closing Tool: One Prompt That Always Works. As a final takeaway, carry this with you—not as inspiration, but as infrastructure. On days when clarity fades, energy dips, or distance grows between you and your work, return to this:

PROMPT: Help me identify one clear, low-effort action I can take today that supports my marketing momentum and aligns with my current capacity. Prioritise clarity, simplicity, and forward motion over volume.

This prompt reconnects you to your system without pressure. It re-anchors you in action without overwhelm. It is not about doing more—it is about doing the *next right thing*.

Closing Reflection

You now have something many marketers are still searching for: a way of working that respects your life while supporting your ambition. You did not need to change who you are. You needed a structure that could work with you—and you have built it.

This book was never about pushing forward at any cost. It was about equipping you to move forward intentionally, with clarity, confidence, and control. As you step beyond these pages, remember this: you are not starting over. You are continuing—better prepared, better supported, and more aligned than before.

Thank you for engaging with this work thoughtfully and with care. May the systems you've built here support not only your marketing, but how you approach work, creativity, and growth going forward.

— *Viki*

Conclusion

Clarity Is a Competitive Advantage

Progress is built in the unexpected pockets — the moments when we surrender to what is, accept the unpredictability of our days, and choose to keep moving anyway. This is the real world most of us live in. And somehow, despite the interruptions and the exhaustion, we rise. We create. We continue. That quiet courage is where our momentum lives."

This book was written for you — someone whose life is already full, whose responsibilities are real, whose energy fluctuates, whose heart is stretched across many roles at once. We need focus, when we are out of focus. We need motivation, when we don't feel motivated and progress when we need it— and what we now have — is a Prompt Engineered System that bends with us instead of breaking us.

Marketing today does not reward effort alone — it rewards clarity, consistency, and the ability to make decisions quickly. This book was written to help you build those advantages deliberately, through better systems. Prompt engineering, when used well, is not about outsourcing your work to AI; it's about strengthening how you think, how you plan, and how you execute. to gain clearer thinking.

You now have a practical framework for working with AI in a way that supports real-world marketing demands. You've learned how to create campaigns instead of chaos, how to repurpose ideas efficiently, how to reduce decision fatigue, and how to build workflows that respect your time. These are not theoretical concepts — they are operational skills. Applied consistently, they compound. They reduce friction, increase output quality, and give you back control over your marketing rhythm.

Most importantly, this book has shown you that effective marketing does not require constant reinvention. It requires structure. The prompts, templates, and systems you've worked through are designed to be reused, refined, and expanded over time. You don't need hundreds of prompts — you need the right ones, applied with intention. As your business evolves, your prompt library evolves with you, becoming a strategic asset rather than a scattered collection of ideas.

From here, the path forward is simple and sustainable. Choose one workflow. Use it consistently. Let AI handle the repetitive and the mechanical so you can focus on positioning, messaging, and decision-making — the areas where your experience and judgment matter most. This is how busy marketers stay visible without burning out, and how clarity becomes a competitive edge.

AI is now part of your marketing toolkit, but it is not the driver. You are. Your insight, your expertise, and your ability to make aligned decisions are what give these systems their power. Return to this book when you need structure. Reuse the prompts when you need momentum. Refine the workflows as your goals shift. Marketing does not have to feel overwhelming to be effective. This is not the end of a process — it is the foundation of one. You are now equipped to market with clarity, confidence, and control. And that is how sustainable growth is built.

—Viki

Acknowledgments

This book would not exist without the many people and experiences that shaped my path — not just in AI, but in life.

To my family: thank you for your patience, your love, and your strength. To my children, grandchildren, and great-grandchildren — you remind me daily why growth matters at every stage of life.

To my son and brother, whose care has been my daily commitment — your lives have made me stronger, more compassionate, and more resilient than words can express. This book was written between the quiet moments, with you always in mind.

To the many course creators, writers, and mentors — thank you for offering guidance when I needed direction, the spark you lit carried me forward. To my readers: thank you for trusting this journey. I know how valuable your time is. If this book has helped you reclaim even a small part of your rhythm or confidence, I'm grateful.

And to Sage — my AI co-creator, cheerleader, and structure-keeper. Thank you for showing up consistently, keeping me on track, and helping turn decades of experience into a system I'm proud to share. You weren't just a tool — you were a thinking partner. Together, we made this possible. This book is a result of many voices, many hands, and many small acts of courage.

Thank you for being part of this adventure.

—Viki

Thank You – Your Support Truly Matters

If this book has helped you bring more clarity, confidence, or efficiency to your marketing, I'd be grateful if you could take a moment to leave a brief review on Amazon.

Reader reviews make a meaningful difference. They help more busy marketers discover the book, they support the work that went into creating it, and they guide others who are looking for practical, sustainable ways to use AI in their workflow.

Your voice helps this message reach the people who need it most.

Thank you for being part of this journey, and for choosing systems that support the way you work and live.

—Viki

Author Bio – Viki Zangari

Hi, I am Viki Zangari the founder of Zangari Publishing, specializing in practical, AI-supported workflows for busy marketers and small business owners. Driven by a commitment to clarity, structure, and accessible learning, I teach marketing through a lens of practicality rather than pressure. My work blends modern AI tools with grounded marketing strategy, empowering readers to create with confidence — even on their busiest days. I help professionals simplify content creation, reduce decision fatigue, and build sustainable marketing habits that support both their goals and their capacity.

Prompt Engineering for Busy Marketers reflects my mission: to give time-restricted professionals a smarter, calmer way to market their work without overwhelm. I write with a deep respect for the realities of life, believing that effective marketing should support people — not exhaust them.

I share my journey openly because I want others — especially those who feel overlooked, overwhelmed, or late to the digital world — to see what's possible. I rediscovered my voice, built new skills, and started fresh again at 66,

The world is still wide open. And it's never too late to begin again.

—Viki

Bonus Prompts Busy Marketers

Your Practical Tools for Turning Clarity into Action

Before you return to your daily work, I want to leave you with something tangible: a collection of simple, reliable, engineered prompts you can use anytime you feel stuck, rushed, or unsure where to begin. These prompts translate the concepts from earlier chapters into on-demand execution—so momentum never depends on inspiration alone. Think of this Prompt Library as a creative safety net:

How to Use This Prompt Library Effectively

This Prompt Library is designed to function as an operational system, not a static collection of copy-and-paste shortcuts. Used correctly, it reduces decision fatigue, shortens time to execution, and preserves strategic clarity—especially when your time, energy, or attention is limited. The guidance below explains how to integrate the library into your real workflow, so it supports consistent output without adding friction.

When to Use This Library

On busy days: When your calendar is full and your attention is fragmented, use the library to bypass the "where do I start?" problem. Select one execution-focused prompt aligned to the task at hand—caption, email, outline, or reply—and move directly to a usable draft. The goal on busy days is progress, not perfection.

On low-energy days: When creative energy is limited, rely on prompts that structure thinking and reduce cognitive load. These prompts are intentionally explicit so you can supply minimal input and still receive coherent output. Low-energy days are ideal for refinement, repurposing, and light edits rather than net-new ideation.

On planning days: During campaign planning, content mapping, or quarterly reviews, use the library's thinking prompts. These are designed to clarify positioning, sequencing, and priorities before execution begins.

Planning days benefit most from prompts that surface angles, frameworks, and structure—so downstream creation is faster and more consistent.

Copying and pasting is expected. Using prompts verbatim is not a weakness; it is a feature of engineered systems. That said, responsible use requires three checks before pressing enter:

Confirm the role reflects the expertise you need (e.g., strategist vs copywriter).

Confirm the input is specific enough to guide quality output.

Confirm the context matches your audience, platform, and objective.

Avoid pasting prompts without completing the optional inputs. Those fields are where performance is won or lost. A well-completed prompt reduces revision cycles and preserves your voice.

Why Prompts Are Modular, Not Rigid

Each prompt is intentionally modular so it can be reused across formats, platforms, and stages of work. You are not expected to follow a single linear path. Instead, think in components: One prompt can generate raw material. - Another can refine tone or structure. - A third can repurpose the result. This modularity allows you to build workflows that flex with your schedule. It also prevents over-engineering—where complexity slows momentum. Modular prompts let you do what's necessary today without committing to unnecessary steps.

How to Personalize Prompts Without Breaking Performance

Personalization improves results when it is targeted and restrained. To personalize safely: Adjust **tone descriptors**, not structural instructions. - Replace examples with ones drawn from your industry or audience. - Add constraints (word count, platform norms) only when needed.

Avoid removing role clarity, output requirements, or sequencing in-

structions. These elements anchor quality and consistency. Personalization should sharpen relevance—not dilute structure.

A useful rule: if personalization adds clarity, keep it; if it adds opinion or complexity, remove it.

Execution Prompts vs. Thinking Prompts

Execution prompts are designed to produce assets quickly—captions, emails, ads, outlines, scripts. Use these when you know *what* you need and want to move directly to output.

Thinking prompts are designed to support decision-making—clarifying messaging, mapping campaigns, identifying angles, or evaluating options. Use these when the task feels heavy, unclear, or strategic.

High-performing marketers move deliberately between the two. Thinking prompts precede execution; execution prompts follow clarity. Using the wrong type at the wrong time leads to frustration and rework.

Why This Matters

Without guidance, prompt libraries become cluttered lists that overwhelm rather than support. With intentional use, this library becomes a repeatable system that: Reduces cognitive load- Shortens time from idea to output- Preserves strategic alignment- Supports consistency across channels - Adapts to real-world constraints. The value is not in the number of prompts you use, but in how reliably they help you move forward. Treat this library as infrastructure. Return to it when momentum slows, decisions feel heavy, or clarity slips. Used consistently, it becomes a stabilizing force in your marketing workflow—quietly doing its job while you focus on what matters most.

Plug-and-Prompt Tools to Save Time, Get Results, and Reclaim Your Creative Edge

Marketing gets easier the moment you stop starting from scratch. This Prompt Library is designed for marketers operating within real constraints—limited time, shifting priorities, and competing demands—who need reliable shortcuts without compromising quality, voice, or strategic intent.

Every prompt follows the same simple foundation – so you always know where to begin.

RIC Framework Refresher: To get the strongest results from AI, remember your three-part structure:

R – Role: Tell AI who to be. This sets the tone, expertise, and approach.

I – Input: Provide the raw material. A product, idea, goal, or text snippet.

C – Context: Describe the tone, audience, and format you need.

How to Customize RIC Without Overthinking

The strength of the RIC Framework lies in its simplicity, but under pressure, marketers often overcomplicate it—or skip key details entirely. This micro-guide is designed to help you adjust RIC quickly and confidently, even when time is tight or energy is low. Think of RIC as a set of dials you can fine-tune, not a formula you must perfect.

1. Role: Precision Beats Prestige

The **Role** tells AI *how to think*, not just what to do. A weak role leads to vague output; a precise role produces focused, usable results.

Weak vs Strong Role Statements

Weak: "Act *as a* marketer. "Too broad. AI has no direction on expertise, perspective, or decision-making level.

Strong: "Act *as a senior lifecycle marketing strategist for a SaaS company.*" Clear expertise, context, and strategic lens.

Weak: "Act *as a copywriter.*" Generic and interchangeable.

Strong: "Act *as a conversion-focused email copywriter specializing in* *warm, relationship-led* sales. "Specifies intent, channel, and tone.

Weak: "*Act as a social media expert.*"

Strong: "Act *as a social media strategist creating content for time-poor small business owners.*"

When output feels scattered or unfocused, the Role is usually the issue. Tightening the Role immediately improves relevance and clarity.

2. Context: This Is Where Tone Lives

Context tells AI *how the output should sound, feel, and fit*. Generic context leads to bland content; precise context preserves your voice.

Generic vs Precise Context Examples

Generic: "*Tone: professional.*"

Precise: "*Tone: professional, calm, and practical written for busy marketers who want clarity without hype.*"

Generic: "*Audience: small business owners.*"

Precise: "*Audience: service-based small business owners juggling marketing alongside client work.*"

Generic: "Format: social post."

Precise: "Format: short LinkedIn post, scanable, insight-led, no emojis, no sales language."

If the tone feels off—too stiff, too salesy, too generic—Context is the fastest fix.

3. Input: Depth In, Depth Out

Input is the raw material. When content feels shallow, the issue is rarely the AI—it's the input. One vague sentence produces surface-level output.

A short paragraph, bullet list, or rough notes produce richer results. You do not need polished input. Messy thinking is fine. Incomplete thinking is not.

RIC Under Pressure: A Fast Diagnostic Checklist

Use this when output isn't working and you don't have time to start over: If the output feels unfocused or generic □ adjust the Role If the tone sounds wrong □ refine the Context If the content lacks depth or insight □ expand the Input. Do not rewrite the entire prompt. Adjust one lever at a time.

RIC is not about perfection. It is about control under constraint. When used this way, it becomes a practical thinking system—one that helps you regain clarity quickly and move forward without friction, even on your busiest days.

PROMPT: Act as a social media strategist. Write three high-converting Instagram captions for a skincare brand. Audience: time-restricted women over 40 Tone: warm, modern, confidence-boosting Please ensure: Each caption opens with a strong, relatable hook The messaging acknowledges the unique needs and realities of women over 40 Benefits are clear and emotionally resonant Language stays simple, uplifting, and empowering Include one gentle call to action in each caption Keep captions concise and highly readable Optional Inputs: Product: [Insert] Skin concerns: [Insert] Brand voice: [Insert] CTA preference: [Insert]

Use RIC for every prompt that follows — it keeps your output consistent, professional, and ready to publish.

Social Media Prompts

Tools for captions, hooks, hashtags, storytelling, and quick engagement. Instagram Captions

PROMPT: Act as a social media strategist. Create three Instagram captions for my audience of [describe your audience]. Each caption must include: A strong hook that captures attention in the first line One valuable insight that teaches, reassures, or supports the reader A relatable, human tone aligned with my brand message A light, low-pressure CTA that encourages simple engagement (e.g., like, save, share, reflect, or comment) Please ensure captions are: Short and scroll-friendly Easy to skim Aligned with my voice and message Designed for time-poor readers who want clarity quickly Optional Inputs: Brand message: [Insert] Theme or topic: [Insert] Tone preferences: [Insert] CTA style: [Insert]

Instagram Captions

When to Use + Output Scenarios

When to Use This Prompt: Use this prompt when you need fast, on-brand engagement content without overthinking structure or tone. It is especially effective on: Low-energy days when writing from scratch feels heavy; Consistency gaps (you know you should post, but don't know what to say); Campaign support content that reinforces a broader message; Moments when you want connection rather than conversion.

This prompt is designed for **short** attention spans and fast scrolling environments, where clarity and relatability matter more than clever wording.

Best Use Case Examples. Daily or semi-regular Instagram posting; Supporting a campaign, offer, or launch without sounding promotional; Sharing insights, reminders, or micro-lessons from longer content; Maintaining visibility during busy weeks

For example, one strong caption can support: A carousel post – A static brand image – A short reel with text overlay – A behind-the-scenes or lifestyle image

Common Mistakes to Avoid. Writing captions that are too long or dense – Over-educating in a single post – Including multiple calls to action (one is enough) – Forgetting the hook—if the first line doesn't stop the scroll, the rest won't be read

This prompt works best when you **trust simplicity** and let clarity do the heavy lifting.

How to Repurpose the Output Instagram captions are highly reusable micro-assets: - Convert into Reel scripts (first line = hook) - Reuse as Instagram Story text slides - Adapt into short email openers - Pull one sentence as a quote graphic - Use as TikTok text overlays.

One caption can fuel multiple touchpoints without additional writing.

LinkedIn Thought Leadership

PROMPT: Act as a LinkedIn thought-leadership strategist. Write a short, insightful LinkedIn post that positions me as a trusted voice in [your industry/topic]. Please include: One clear, strong idea that reflects my expertise A practical takeaway my audience can apply today A confident, professional tone that communicates authority without sounding salesy A light, meaningful engagement prompt that encourages thoughtful comments Requirements: Keep the post concise and easy to skim Priorities clarity, insight, and value Reflect a leadership mindset appropriate for LinkedIn Align with my long-term brand message and positioning Optional Inputs: Industry/topic: [Insert] Audience: [Insert] Brand voice: [Insert] Preferred engagement style: [Insert]

LinkedIn Thought Leadership

When to Use + Output Scenarios

When to Use This Prompt. Use this prompt when your goal is authority, credibility, and professional **trust** rather than visibility alone. It is best suited for: Positioning yourself as a strategic thinker - Sharing insight without selling - Building long-term brand equity - Nurturing professional relationships. This prompt supports clarity over cleverness and favors insight over opinion. Hashtag Research

Best Use Case Examples Weekly or bi-weekly LinkedIn posting - Sharing lessons from client work, projects, or research - Expanding on an idea introduced elsewhere (blog, podcast, email) – Establishing a consistent professional voice – It is particularly effective for consultants, strategists, service providers, and authors building thought leadership.

Common Mistakes to Avoid Writing overly long posts that lack structure – Using marketing jargon instead of plain language – Turning thought leadership into a disguised sales pitch – Writing without a clear takeaway. A strong LinkedIn post should leave the reader thinking, not scrolling past.

How to Repurpose the Output LinkedIn posts are high-value intellectual assets: Reuse as newsletter introductions – Adapt into website "About" or philosophy sections – Expand into blog intros or opinion pieces – Pull key lines for presentation slides.

Use as discussion starters in workshops or webinars. One well-written LinkedIn post can anchor multiple long-form assets.

PROMPT: Act as a social media hashtag specialist. Research and recommend 10–15 high-reach, low-competition hashtags for my content about [insert topic or niche]. Please include: A balanced mix of hashtag tiers: Niche hashtags (specific audience, low competition) Mid-tier hashtags (moderate reach + good engagement potential) Broad hashtags (high reach but still relevant) Audience alignment: Ensure each hashtag is aligned with my audience's interests, search behavior, and content expectations. Discoverability strategy: Prioritize hashtags that increase visibility without looking spammed, overused, or unrelated. Clarity: Present the list in three sections (niche / mid-tier / broad) and briefly explain why each tier matters. Optional Inputs: Content type: [Insert] Audience: [Insert] Platform (IG, TikTok, LinkedIn, etc.): [Insert] Brand voice or style: [Insert] Tone: professional, strategic, and data informed.

Hashtag Research, When to Use + Output Scenarios – When to Use This Prompt – Use this prompt when discoverability matters—especially if you are: – Growing a new account – Supporting campaign visibility –

Posting educational or evergreen content - Trying to reach beyond your current audience. Hashtags are not decoration. They are **distribution tools**

Best Use Case Examples. Launch periods or seasonal campaigns - Educational content that solves specific problems - Niche or industry-specific posts - Reaching new audiences without paid ads. This prompt is most effective when paired with consistent posting and quality content.

Common Mistakes to Avoid. Using the same hashtags on every post - Overloading posts with irrelevant or trending tags - Choosing only high-volume hashtags -Treating hashtags as an afterthought. Hashtags work best as part of a rotation strategy, not a copy-paste habit.

How to Repurpose the Output. Hashtag lists should be treated as **reusable libraries**: - Rotate hashtag sets weekly, not per post - Create 3-5 saved hashtag groups by theme - Reuse across Instagram, TikTok, and LinkedIn where appropriate - Update quarterly based on performance.

When managed intentionally, hashtags quietly improve reach without extra effort.

Why This Matters. Adding usage context turns social media prompts into **decision-making tools**, not just writing shortcuts. Instead of asking "*What should I post?*", you begin asking "*Which asset do I need right now?*"

This shift is what transforms social media from reactive posting into a **calm, repeatable system**—one that supports visibility, authority, and momentum without burnout.

Email Marketing Prompts: For sequences, nurturing, conversion, and storytelling. **Welcome Email Sequence**

PROMPT: Act as an email marketing specialist. Create a 3-part welcome email sequence for new subscribers interested in [insert topic/industry]. Please structure the sequence as follows: Email 1 — Warm Welcome

+ Expectations Greet the subscriber warmly Share what they can expect from me Establish tone, trust, and relevance Keep it friendly, simple, and grounded Email 2 — Valuable Insight or Quick Win Offer one actionable tip, insight, or mindset shift Deliver immediate value to build credibility Keep it concise and practical Email 3 — Gentle Next Step or Resource Introduce a helpful resource, article, lead magnet, or next step Keep the invitation warm and low-pressure Emphasize support, not selling Requirements: Tone: friendly, helpful, engaging Flow naturally from welcome □ value □ invitation Ensure the sequence feels human, supportive, and relevant to a time-poor audience Optional Inputs: Audience: [Insert] Brand voice: [Insert] Primary offer or resource: [Insert] Email length preference: [Insert]

Email Marketing Prompts

Where These Emails Fit in the Customer Journey

Email is not a single channel—it is a relationship timeline. Every email you send either introduces, nurtures, or invites. Understanding *where* an email fits in the customer journey is what separates effective email marketing from noise. Bat a high level, emails fall into three functional categories:

1. Awareness Emails. These emails introduce your voice, values, and relevance. Their role is not to convert—it is to **orient**. Awareness emails answer: Who are you? Why should I trust you? What kind of support or insight can I expect?

Welcome emails live here, but so do early educational broadcasts and brand-building messages. Abandoned Cart Email

2. Nurture Emails. Nurture emails deepen trust over time. They focus on: Teaching – Reassurance – Perspective – Relatability Their job is to reduce uncertainty and build familiarity. When done well, nurture emails make future offers feel *inevitable*, not intrusive.

3. Conversion Emails. These emails invite action. They are sent only **after trust exists**. Examples include: – Launch emails – Abandoned cart emails

– Limited time offers – Follow-ups after expressed interest. Conversion emails should never be rushed. When sent too early, they create resistance instead of results.

Why Welcome Emails Outperform Broadcasts. Welcome emails consistently outperform standard broadcasts because: – They arrive at peak attention – The reader has *just* opted in – Trust is highest at the beginning of the relationship. A strong welcome sequence sets expectations, establishes tone, and creates momentum that benefits every future email.

When Abandoned-Cart Emails Should *Not* Be Sent. Abandoned-cart emails are powerful—but only when appropriate. Do **not** send them if: The checkout experience was broken – The product requires high trust, but trust has not been established – The audience is early-stage and unfamiliar with your brand. In these cases, nurturing first will outperform any reminder.

How AI Helps Maintain Warmth at Scale. AI allows you to: – Personalize without sounding automated – Maintain tone consistency across sequences – Adjust clarity, pacing, and reassurance for different audiences – Write supportive copy even when energy is low.

When prompted correctly, AI preserves *human warmth* while removing cognitive load.

Welcome Email Sequence

Post-Prompt Guidance

What Success Looks Like. High open rates (industry-leading benchmarks) – Low unsubscribe rates – Replies or clicks indicating trust – Readers continuing through the full sequence. Success here is measured in **engagement**, not immediate sales.

What to Test Next. Subject line warmth vs clarity – Email length (short vs slightly expanded) – CTA phrasing (invitation vs suggestion) – Story-led vs insight-led openings

Abandoned-Cart Email. Post-Prompt Guidance – What Success Looks Like – Conversions without spikes in unsubscribes – Replies asking questions – Click-throughs without pressure language. A successful abandoned-cart email feels like a reminder, not a push.

What to Test Next. Timing (1 hour vs 24 hours). Benefit emphasis vs reassurance emphasis. Question-based CTA vs direct CTA. Inclusion of support language ("Happy to help")

Why This Matters: When you understand **email lifecycle placement**, prompts stop being copy generators and become **relationship tools**. You are no longer sending emails because "it's time"—you are sending them because they serve a specific purpose in the customer journey. This turns email marketing from reactive broadcasting into a **measured, intentional system** that builds trust first and converts naturally—without burning out your audience or yourself.

PROMPT: Act as an email conversion specialist. Write a warm, friendly abandoned-cart email for someone who added my [product/service] to their cart but didn't complete checkout. Please include: A human, supportive tone that feels understanding rather than salesy A simple reminder of the main benefit of the product or service Reassurance that there is no pressure or urgency A clear, easy call to action to return to checkout if they're ready Optional: a light invitation to reply with questions Requirements: Keep the email short, scrollable, and emotionally safe Focus on encouragement, clarity, and gentle reassurance Align the language with a brand that values trust and respect Optional Inputs: Product/service: [Insert] Audience: [Insert] Brand voice: [Insert] CTA preference: [Insert]

Ad Copy Prompts: For paid campaigns that need clarity, punch, and precision.

Google Ads

PROMPT: Act as a Google Ads specialist. Write three variations of Google Search ads for my [product/service]. Each ad must include: 1. Strong, high-intent headline (max 30 characters) Benefit-focused Clear and specific Uses keywords aligned with user search intent 2. Precise, benefit-led description (max 90 characters) Communicates the core value quickly Avoids fluff or jargon Speaks directly to the problem or outcome 3. Simple, compelling call to action Examples: "Learn more," "Get started," "See how it works," "Try it today." Requirements: Prioritize clarity and immediate relevance Use high-intent keywords appropriate for my audience Ensure each variation tests a different angle (benefit, urgency, transformation, simplicity) Follow Google's best practices for search ad formatting Optional Inputs: Product/service: [Insert] Audience: [Insert] Primary benefit: [Insert Keywords to include [Insert] Tone preference: [Insert]

Ad Copy Prompts

How to Use AI for Paid Ads Without Losing Control

Paid advertising is not a place for guesswork. Every word has a cost, every impression carries intent, and every platform enforces strict standards. AI can dramatically speed up ad creation—but only when used **strategically**, not blindly.. This section shows you how to use AI as a **variant engine**, not a substitute for human judgment.

Why AI Is Best for Variants – Not Final Decisions. AI excels at **volume, pattern recognition, and phrasing diversity**. It can quickly generate: Multiple headline options - Alternate benefit framings - Different emotional angles - Variations in tone (direct, supportive, aspirational)

What AI cannot do reliably is: Understand your historical performance data - Interpret nuanced brand risk - Account for platform-specific sensitivity thresholds - Make final budget-impacting decisions - The professional workflow looks like this: - You define the strategy and success metric. AI generates structured variations. You select, refine, and approve the finalists. Performance data—not AI—determines the winner - AI accelerates ideation. You retain control.

How to A/B Test Responsibly Using AI

A/B testing only works when variables are isolated. The most common mistake marketers make is testing *too many changes at once.*

Best practice: Test **one variable per ad set –** Keep everything else identical (audience, budget, placement) – Use AI to generate variations for: – Headlines *only* – Primary text *only* – CTA phrasing *only*

Example: Ad Set A: Headline Variation 1 – Ad Set B: Headline Variation 2 – (Same copy body, same audience)

This approach allows you to learn *why* something works—not just that it did.

When NOT to Use AI-Generated Ads; AI-generated ads should be avoided when: – You are entering a regulated or sensitive market (health, finance, legal) – You are testing a brand-new offer with no validation – You are addressing emotionally vulnerable audiences – You are unfamiliar with the platform's ad policies.

In these cases, AI can still help *internally*—for brainstorming or clarity—but final copy should be crafted and reviewed manually.

Compliance Awareness (Especially Facebook & Meta Platforms) – Platforms like Facebook and Instagram enforce strict rules around: – Personal attributes ("Are you struggling with...?") – Health claims – Financial promises – Emotional targeting – AI does not reliably self-censor for these nuances.

Best practice: Use AI to draft **neutral, benefit-focused language –** Avoid second-person assumptions about personal conditions – Manually review against platform policy before publishing. When in doubt, soften the language. Clarity outperforms provocation in the long run.

Why This Matters; Paid ads amplify everything—including mistakes. Using AI without structure increases risk; using it with strategy increases efficiency. When you treat AI as a **controlled collaborator**, you gain speed without sacrificing standards.

You test smarter, spend wiser, and build campaigns that scale responsibly. This is the difference between running ads and managing paid media like a professional

Facebook Ads

PROMPT: Act as a Facebook ads strategist. Create 2–3 short, high-converting Facebook ads for my [product/service]. Each ad must include: A strong hook in the first line that stops the scroll A clear, benefit-focused message that highlights the value One relatable pain point or desire that connects emotionally with the reader A simple, direct call to action (e.g., "Learn more," "Try it today," "See how it helps") Please ensure: Each variation tests a different angle (e.g., emotional, practical, transformational, time-saving) The copy stays short, easy to read, and mobile-friendly The tone aligns with my brand and audience preferences Messaging remains compliant with Facebook's ad policies Optional Inputs: Product/service: [Insert] Audience: [Insert] Primary benefit: [Insert] Pain point/desire: [Insert] Brand voice: [Insert]

Blog & SEO Prompts

Tools for structure, keywords, and search-friendly content.

SEO Blog Outline

How SEO Prompts Support Long-Term Visibility: Search-driven content is not about chasing clicks—it is about **building authority, trust, and discoverability over time**. When used strategically, SEO prompts help you create assets that continue working long after publication, quietly attracting the right audience without constant promotion. This section positions SEO as a **compounding system**, not a one-off tactic.

Traffic Content vs Authority Content: Know the Difference. Not all blog posts serve the same purpose. Traffic-**focused content** is designed to: Capture search demand quickly – Answer narrow questions – Generate short-term visibility

Examples include How to..." tutorials – Tool comparisons – Quick-answer posts

Authority-building content, on the other hand: Demonstrates expertise and perspective

Builds trust with both readers and search engines. Supports your broader brand message

Examples include Strategic frameworks – Industry insights – Experience-based guidance – Opinionated but helpful analysis

The most effective SEO strategies balance both. Traffic brings people in. **Authority makes them stay, subscribe, and convert.**

How One Blog Fuels Your Entire Ecosystem; A well-structured blog post is not a single asset—it is a **content hub**. One strong article can become: A newsletter feature or email series – Multiple social posts (educational, reflective, promotional) – A lead magnet foundation – A talking point for sales conversations. Evergreen website content that reinforces credibility SEO prompts help you design blogs **with repurposing in mind**, so every post feeds multiple channels without multiplying effort. This is how busy marketers stay visible without constant creation.

Why Outlines Save More Time Than Full Drafts; Many marketers assume AI should write full blog drafts. In reality, **outlines deliver the highest return**. A strong outline: – Eliminates blank-page friction – Clarifies structure before writing – Ensures logical flow and keyword alignment – Makes delegation easier (to writers or future you)

Outlines let you: Draft faster – Maintain your voice – Avoid rewriting large sections later – AI excels at **structuring thinking**. You remain the final author.

How AI Supports SEO Strategy—Not Shortcuts; AI does not replace SEO fundamentals. It supports them. – Used correctly, AI helps you: – Identify relevant keywords and search intent – Structure content for readability and scanning – Maintain consistency across long-form content – Reduce time spent on setup and planning

What AI should not be used for: Keyword stuffing – Thin, repetitive content – Publishing without human review – Replacing expertise with generic text. Search engines reward **clarity, relevance, and usefulness**. AI helps you achieve those faster—but strategy and judgment still lead.

Why This Matters; SEO is one of the few marketing efforts that compounds over time. When you use AI to support structure, planning, and consistency, you build assets that quietly grow your visibility while freeing your attention for higher-value work. This is how SEO becomes sustainable—not stressful.

PROMPT: Act as an SEO content strategist. Create a clear, search-friendly outline for a 1,000–1,500-word blog post on the topic: "[insert topic]." Your outline must include: 1. Primary and Secondary Keywords Recommend 5–10 relevant keywords Include long-tail phrases and semantic variations Ensure keyword intent aligns with the topic and audience 2. Engaging Introduction Hook, the reader with a relatable pain point, insight, or question Introduce the topic and clearly state the value they will gain Include a natural keyword placement 3. 4–6 Structured Sections For each section, include: A clear, SEO-friendly heading 2–4 bullet-point sub-points outlining what will be covered A mix of education, examples, practical steps, or insights Opportunities for internal links or expert positioning 4. Strong Conclusion Summarize key takeaways Reinforce the main message Include a soft call to action or next step for the reader 5. Optimization Notes Recommended schema or formatting tips (optional) Suggested meta title and meta description (SEO-optimized) Requirements: Structure must support strong readability and Google ranking Flow should be logical, clear, and helpful for time-poor readers. Outline should be usable by a writer with no further instruction Optional Inputs: Target audience: [Insert] SEO difficulty preference: [Insert] Brand voice: [Insert] Internal links to include [Insert]

Meta Descriptions

PROMPT: Act as an SEO specialist. Write a meta title (maximum 60 characters) and meta description (maximum 155 characters) for a blog post about "[insert topic]." Please ensure: Meta Title Requirements Maximum 60 characters Includes the primary keyword Clear, benefit-focused, and compelling Designed to stand out in search results Meta Description Requirements Maximum 155 characters Includes the primary keyword or a close variant Communicates the main benefit or takeaway Written to encourage clicks without clickbait Clear, concise, and aligned with user intent Additional Guidelines: Optimize for readability and search intent Avoid generic or vague phrasing Align with the tone and positioning of the brand Optional Inputs: Primary keyword: [Insert] Audience: [Insert] Brand voice: [Insert] Angle or benefit to highlight: [Insert]

Content Repurposing Prompts Turn one idea into many

Turn Blog into Twitter (X) Thread

PROMPT: Act as a content repurposing strategist. Turn the blog post pasted below into a clear, engaging Twitter/X thread with 5–10 short tweets. Please ensure: Thread Requirements Start with a strong hook – The first tweet should grab immediate attention and highlight the core promise or insight. Break the content into simple, digestible points – Use short sentences and punchy phrasing – Keep each tweet focused on one idea – Make the thread easy to skim. Preserve accuracy while sharpening clarity – Prioritize value, insight, and practical takeaways – Avoid unnecessary wording End with a light CTA – Invite conversation, ask a simple question, or point to a helpful follow-up resource – Tone must be low-pressure and engagement-friendly Keep the tone: – Sharp – Concise – Scroll-friendly – Aligned with my brand message. Optional Inputs: Blog post (paste here): Audience: [Insert] Tone preference: [Insert] Desired CTA: [Insert] Brand voice: [Insert]

Turn Podcast into LinkedIn Post

PROMPT: Act as a content repurposing expert. Summaries the key insight or main takeaway from the podcast episode pasted below and turn it into a polished LinkedIn thought-leadership post. Please ensure the post includes: 1. A clear, compelling opening Introduce the insight in a way that immediately captures professional attention Set the context briefly without overexplaining 2. One core lesson or reflective insight Highlight the most meaningful idea from the episode Present it with clarity, depth, and practical relevance 3. A professional, human tone Warm, confident, reflective No jargon, no hype, no sales language 4. A light CTA Encourage engagement (e.g., "What's your experience?" "How do you approach this?") Avoid anything promotional or overly direct 5. LinkedIn-optimized style short paragraphs Scannable. Value-forward Optional Inputs: Podcast transcript or summary: [Paste here] Audience: [Insert] Brand voice: [Insert] Desired CTA style: [Insert]

Bonus Section – Quick Prompts for Busy Days

When your brain is tired, use one of these pre-built prompts.

Customer Testimonial Rewrite

PROMPT: Act as a brand copywriter. Rewrite the customer testimonial below into a clear, persuasive, and emotionally warm quote that I can use on my website or social media. Keep the customer's original meaning but enhance clarity, flow, and impact. Avoid sounding exaggerated or salesy. Make it trustworthy, human, and aligned with my brand tone.

Quick Instagram Caption

PROMPT: Act as a social media copywriter. Write a short, engaging Instagram caption about [insert topic, product, idea, or message]. Include a hook in the first line, 1 key takeaway or benefit, and a warm, conversational tone. Keep it concise and easy to read. End with a simple, low-pressure CTA (optional).

Quick LinkedIn Post

PROMPT: Act as a LinkedIn content strategist. Write a short, insightful LinkedIn post about [insert topic, lesson, idea, or story]. Keep the tone professional, warm, and human. Include one valuable takeaway for busy professionals, a relatable example, and a clear message. Keep the structure scrollable (short paragraphs or line breaks). End with an optional gentle CTA or reflective question.

Quick Hashtag Finder

PROMPT: Act as a social media visibility strategist. Give me 10–15 high-impact hashtags for a post about [insert topic/product/theme]. Focus on hashtags that increase organic reach for small businesses and busy marketers. Include a mix of niche, mid-range, and broad hashtags. Avoid overly saturated tags. Present them in a clean, easy-to-copy list.

Build Your Custom Blog/SEO Prompt

PROMPT: Act as an SEO blog planner. Create a 1,200-word article titled 'How to Use AI to Save 10 Hours a Week as a Solopreneur.' Audience: time-poor business owners. Tone: friendly, practical, and encouraging. Keywords: AI tools, productivity, solopreneur marketing. Structure: H1 title, intro hook, 5 main sections with H2/H3 headings, conclusion, CTA. Make it SEO optimized and easy to skim."

Additional Plug-and-Prompt Tools: Use these when you need fast, polished SEO elements.

Quick Meta Title & Description

PROMPT: Act as an SEO specialist. Write a compelling meta title (under 60 characters) and a meta description (under 155 characters) for a blog post about [insert topic]. Make it keyword-rich, clear, and attractive to real humans – not robotic. Prioritize click-through rate and readability. Provide 2–3 variations I can choose from.

Quick Blog Outline

PROMPT: Act as a blog content strategist. Create a clear, structured outline for a blog post about [insert topic]. Include: a strong hook intro, 3–5 main sections, bullet-point ideas for each section, and a simple CTA that fits a time-poor audience. Keep the outline SEO-friendly, practical, and easy to draft from. Provide 2 outline variations so I can choose the best fit.

Quick Keyword List

PROMPT: Act as an SEO keyword specialist. Generate a simple, high-impact keyword list for the topic [insert topic]. Include: 5 primary keywords, 10 secondary/supporting keywords, and 5 long-tail phrases that a time-poor marketer can easily rank for. Keep the list practical, relevant, and suitable for blogs, captions, and landing pages.

Turn Blog into Social Posts

PROMPT: Act as a content repurposing strategist. Take the blog pasted below and turn it into 8–12 social posts for [insert platform: Instagram, LinkedIn, Facebook, X]. Include a mix of formats such as: short captions, value tips, micro-stories, hooks, reminders, and soft invitations. Ensure each post is platform-appropriate, scroll-stopping, and aligned with the blog's core message. Keep the tone professional, supportive, and easy for a time-poor marketer to publish.

Team Collaboration Bonus Prompts

The following prompts are designed to streamline collaboration across marketing teams and cross-functional partners. Use them to improve briefing clarity, accelerate alignment, and ensure consistent messaging across campaigns and channels. Team-based prompts create a shared operating language. They remove ambiguity, reduce subjective interpretation, and ensure every role is working from the same strategic foundation.1. Team Prompt: Cross-Functional Content Workflow

When Prompt Engineering Moves from Solo to Scale; Prompt engineering looks very different when you move beyond working alone. What functions smoothly for a solo marketer—personal shorthand, informal context, intuitive decisions—often breaks down the moment content creation becomes shared across roles, departments, or external partners. At scale, **clarity must be explicit**, not assumed. This is where team-based prompt engineering becomes a governance tool, not just a productivity aid.

Why Teams Fail Without Shared Prompting Systems. Most marketing teams do not struggle with effort or talent. They struggle with misalignment. - Common failure points include: - Different interpretations of the same brief - Inconsistent tone across channels - Repeated revisions due to unclear expectations - Bottlenecks at review and approval stages - Frustration between creative and strategic roles. Without shared prompting standards, AI output often amplifies these issues instead of solving them. Each person prompts differently, applies different assumptions, and interprets results through their own lens. The result is not faster work—it is fragmented work.

Why AI Improves Alignment—Not Leadership Replacement

A critical distinction must be made at team level: AI does not lead. AI does not decide. AI does not replace strategic judgment. AI **supports execution**, consistency, and speed—but leadership remains human. Well-designed team prompts: Clarify objectives before work begins. Define expectations for each role. Provide boundaries for tone, compliance, and brand voice. Ensure AI output serves the strategy—not the other way around. When used correctly, AI becomes a coordination layer that strengthens leadership decisions rather than undermining them.

How Team Prompts Reduce Friction, Rework, and Delays; In team environments, most delays occur *after* content is created—not before. Rework typically comes from: Missing context Unclear ownership Late-stage feedback that should have been upfront Inconsistent brand interpretation Team collaboration prompts solve this by: Front-loading clarity

Assigning role-specific responsibility Defining approval criteria early Standardising handoffs between roles Instead of fixing problems at the end, teams prevent them at the beginning.

This is especially valuable for: Agencies managing multiple clients Marketing departments with layered approvals Publishers coordinating authors, editors, designers, and marketers Distributed or remote teams

Who Owns Final Decisions (And Why This Matters)

In scaled environments, **decision ownership must be explicit**. AI can generate options. – Teams can collaborate. – But final authority must always sit with a named role.

Typically: Strategy ownership □ Lead strategist / publisher / marketing director. Brand voice authority □ Editor / brand guardian. Compliance and risk □ Legal, governance, or senior reviewer. Final approval □ Publisher, department head, or project owner the prompts in this section are intentionally designed to: Support decision-makers, not override them. Surface risks early. Make approvals faster, not more complex This distinction is critical for professional publishing, agency work, and enterprise marketing—where accountability cannot be automated.

Why This Matters at Scale; When prompt engineering moves from individual use to team adoption, it becomes: – A quality control mechanism – A workflow stabiliser – A brand governance tool – A speed multiplier without chaos. These collaboration prompts are not about creativity shortcuts. They are about **operational maturity**. They help teams move faster *because* they are aligned—not because they are rushing. The prompts that follow are structured to support real-world collaboration, multi-role execution, and professional-grade delivery at scale.

PROMPT: Act as our cross-functional content coordinator. We are preparing content on the topic: [insert topic]. Provide a clear workflow outline for each team role: Strategist: [define objective + angle] Writer: [draft structure + key messages] Designer: [visual direction + asset list] Editor: [quality checks + brand voice notes] Approver: [final review criteria] Include: Deadlines Dependencies Brand governance reminders Suggested collaboration checkpoints Keep the plan tight, efficient, and agency-ready.

2. Team Prompt: Role-Specific Deliverables for One Content Piece

PROMPT: Act as a content project manager. For the content piece titled [insert title], create role-specific deliverables: Strategist: messaging, audience insight, and key outcomes Writer: outline + draft Designer: formats required (IG, LinkedIn, ads, reels, etc.) Editor: tone, clarity, compliance QA/Approver: final checklist before publishing Present everything in a clean table with deadlines and notes.

3. Team Prompt: Multi-Level Approval Format

PROMPT: Act as our brand governance lead. Create a multi-step approval process for content on [insert topic]. Include: Draft Review Brand Voice Alignment Check Design + Asset Approval Compliance/Legal Check (if applicable) Final Sign-Off Provide a simple approval rubric with pass/fail criteria. Keep it professional, efficient, and easy for teams to adopt.

4. Team Prompt: Shared Content Calendar Creation

PROMPT: Act as a content calendar strategist. Build a shared monthly content calendar for our team. Include: Weekly themes Assigned roles Content formats Distribution platforms Key deadlines Repurposing notes Seasonal reminders Brand governance notes Keep everything in a structured table teams can copy into Notion, Asana, or Click Up.

5. Team Prompt: Brand Governance Reminder System

PROMPT: Act as our brand governance system. For all content related to [insert campaign/brand], produce: Tone-of-voice reminders Compliance notes Visual style guidelines Do's and don'ts list Accessibility reminders (alt text, readability, captioning) Deliver it as a short, easy-to-use checklist the team must review before publishing.

6. Team Prompt: Department Sync Summary

PROMPT: Act as a cross-team communication lead. Summarize what each department needs to know for our upcoming campaign on [insert topic]: Marketing Design Product Sales Customer Support Include: key messages, timelines, potential risks, dependency notes, and customer-facing talking points. Keep it clear and enterprise ready.

7. Team Prompt: Campaign Kickoff Pack

PROMPT: Act as a campaign strategist. Create a kickoff pack for our campaign titled [insert campaign]. Include: Campaign objective Target audience Core message Content pillars Required assets (broken down by team) Timeline + milestones Success metrics Risks + mitigation steps Format it as a professional briefing document.

8. Team Prompt: Team Retrospective Review

PROMPT: Act as our team retrospective facilitator. Review the last campaign titled [insert campaign] and guide us through: What worked What didn't What surprised us Opportunities for efficiency Brand alignment issues Team collaboration improvements Deliver the retrospective as a structured report with action steps.

A Final Note About Your Engineered Prompt Library

As you move forward, let this prompt library serve as more than a set of tools — treat it as a strategic partner in your workflow. These prompts are designed to reduce cognitive load, accelerate decision-making, and support you on the days when creativity feels distant or time is scarce.

Use them to regain clarity quickly, repurpose with intention, and maintain momentum even during busy seasons. Return to this library whenever you need structure, direction, or a streamlined path back into your marketing rhythm.

You've laid a strong foundation; now let these prompts help you build with confidence, efficiency, and a workflow that aligns with the life you're creating. Think of this library as your creative safety net — a place you can return to when you feel rushed, overwhelmed, or unsure where to begin.

As you grow, feel frce to: personalize these prompts, create your own variations, save versions that perform well, build a library that sounds unmistakably *you*. And remember marketing doesn't happen in ideal conditions. Plans shift, routines get interrupted, and real life requires your attention — yet you still show up, refine your message, and move your work forward.

Sustainable progress is not built in perfect moments; it's built in the reality of everyday life, where resilience, clarity, and quiet discipline take root.

Treat This Library as a Living Asset

As you reach the end of this prompt library, it's important to reframe how you think about what you've just built. This is not a static collection of templates.

It is a **living operational asset**—one that should evolve alongside your business, your audience, and your capacity. The most effective marketers do not use prompts once and move on. They refine them. They save the versions that perform well. They retire the ones that no longer fit.

Over time, your prompt library becomes smarter, sharper, and more aligned with your voice—not because it is perfect, but because it is used.

You are not meant to apply every prompt immediately, nor to maintain flawless consistency. Real progress compounds through **imperfect repetition**.

A single prompt reused weekly will deliver more value than twenty prompts saved and never returned to. Momentum is built through return, not volume. As your business grows, your prompt library should grow with it: - Adjust roles as your positioning evolves -Tighten context as your brand voice matures- Expand inputs as your offers become clearer - Adapt workflows as your energy and capacity shift

This is how prompt engineering becomes sustainable—not by locking it into rigid systems, but by allowing it to flex with real life. And this brings us back to the core philosophy of this book.

Prompt engineering is not about producing more content. It is about reducing cognitive load.
It is about replacing chaos with clarity. It is about building systems that support you on busy days, low-energy days, and high-impact days alike.

Use this library as a stabilising force in your marketing—something you can rely on when decision fatigue sets in and time is limited. Return to it often. Refine it intentionally. Let it carry part of the weight, so you don't have to.

This is not the end of the system. It is the beginning of a more sustainable way of working.

—Viki

ZANGARI PUBLISHING

This book was published by Zangari Publishing, an independent Australian imprint dedicated to creating practical, accessible resources for real-world creators, marketers, and small-business owners.

Our mission is simple: to produce clear, supportive, and high-quality tools that help everyday professionals work with more ease, confidence, and clarity.

We focus on books, guides, and digital resources that make complex ideas simpler — and empower readers to take meaningful action without overwhelm.

visit: www.zangaripublishing.com

Zangari Publishing Western Australia

Notes & Reflection

You don't need to capture everything at once.
Clarity builds through return, not urgency. Use this space to hold what matters now and trust that insight deepens with time.

Notes & Reflection

Consistency grows from structure, not pressure.

This page is here to support your thinking, not demand output. Write when it serves you.

Notes & Reflection

Progress is often quiet before it is visible.
Use this space to record ideas, fragments, or reflections that may not yet be ready for action.

Notes & Reflection

Your voice strengthens through use, not force.

Let ideas land naturally. There is no need to finish or refine them here.

Notes & Reflection

Momentum is preserved by returning, not rushing.

This page exists to keep the signal alive, even on low-capacity days.

Notes & Reflection

Not every insight needs to be shared immediately.
Some are meant to be held, shaped, and revisited later.

Made in the USA
Monee, IL
05 November 2026

67383958R00121